T0340477

The Hormone Balance Handbook

To the loves of my life James and Sebby.

For every woman who feels out of balance,
but doesn't know where to start, I wrote this for you.

The Hormone Balance Handbook

Understand your hormones, take control of your cycle and nourish your whole body

Jessica Shand

Thorsons

Contents

Introduction

Planting seeds takes patience and time, but it's always worth it come summer.

My goal in writing this book is to change the way you think about and act on your health. I want it to encourage you to honour your hormones and celebrate your cycles, whether you are at the start of your hormonal journey, right in the middle or feeling like it could be coming to its end.

By sharing what I've encountered in my personal hormonal journey, what I've tirelessly studied and learned (from both a Western and Eastern nutritional perspective) and, most importantly, experienced from working with real women with real hormonal problems, I will give you solid, nurturing advice and science-backed information to help you nourish yourself to better hormonal health, one meal at a time. My intention is that this book will become a reliable resource that cuts through the complexities, and that you will keep referring to it as you move through the inevitable hormonal ebbs and flows, so that it will always bring you back into balance, in both body and mind.

My Hopes

I hope these pages become crumpled and worn and that there are many, many splatters on them. I hope this is the book you will give to your best friend when she's going through a stressful time or trying to fall pregnant, your daughter when she starts her period or your niece when she's suffering with PMS and breakouts, and that it will help their hormones as it helps yours.

I hope this book makes you realise that while it can be interesting to know the intricate details of exactly how each chemical messenger in your body operates, it's not entirely necessary or helpful, and you don't really need to know all that stuff in order to make a tangible difference.

And I hope this book also empowers you to take ownership of your health and harness your hormones in the most positive and proactive way possible. You know your body better than anyone – it's time to take back control.

Healing the Whole of You

I feel the confusion around hormones over the last few decades has stemmed from an initial lack of education (arguably, still very much dependent on where you go for advice), followed by a universal overcomplication of female health. And it's this combination that's made the topic feel scary and the concept of reaching 'hormone balance' seem unattainable, even faddy at times. It can feel hard to know where to start, and so often many of us either don't bother at all, spend lots of money on self-prescribed supplements or get sucked into trends that can drive imbalances further, making hormone symptoms ten times worse

and leaving us more fed up, confused and depleted than ever.

Don't get me wrong, hormones are complex – there's no denying it – but *how* you manipulate them in order to move the needle in your hormone-balancing journey doesn't have to be. And that's what my mission is centred around: helping you to nourish the key systems in your body and to eat in alignment with your cycle.

I'm going to show you how to make your hormones work for you (and stop them working against you), using food as medicine to encourage your body to do what it's designed to do: to heal and be in balance. And by looking at each of the key body systems and the most common symptoms/areas of concern that I'm presented with in clinic, I'm going to help you achieve hormonal harmony, using specific nourishing recipes and rituals and a simple guide on how to start cycle syncing.

Nothing about our hormones can be looked at in isolation – gaining hormone balance is *essential* to just about every aspect of a woman's health and wellbeing. Which is why the approach in this book is transformative and is the only strategy I know and have experienced first-hand that can make a long-lasting difference.

My Hormonal Health Journey

This is the book I wish I'd read when I started my period and when I was going through my own multi-faceted version of hormone imbalances.

For me, these imbalances manifested during my late teens into my twenties, when I struggled with my changing body image, started overexercising and intentionally undereating. I was punishing my body by exhausting myself at the gym, skipping meals, then bingeing when I felt weak and lost my willpower. I was scared to leave my flat because I was riddled with anxiety and would have panic attacks when I bumped into people I knew on the street. Most significantly for me at the time, I desperately wanted to fall pregnant and found out my polycystic ovary syndrome (PCOS) and thyroid were preventing me from reaching full hormonal health to create new life. It's easy to see now, after years of nourishing my body and committing myself to healing and optimising my health, that my toxic lifestyle habits bred from toxic diet culture and the way I treated my highly hormonally imbalanced body meant my hormones were not going to allow me to conceive – it was too unsafe.

My challenges left me feeling lost and frustrated that I was having to learn things about my hormones and body that should've been taught at school. But I channelled my frustration into immersing myself into the world of holistic health and the more I soaked-up and introduced to my lifestyle the easier it was becoming to stop hating on my body and start helping it; I was quite literally feeding and loving myself better and my mind was blown. I felt like I'd finally woken up to something I had access to all along, real food. I had become so used to feeling less than OK that feeling amazing felt foreign to me.

I decided to take my love for nutrition and enrolled at The College of Naturopathic Medicine where I studied for four years. First as a health coach followed by three further years in Naturopathic Nutritional therapy allowing me to become the registered nutritionist I am today.

This changed my life and has given me the privilege of helping others change theirs too, which is now my life's work. Above all, it has helped me learn how to nourish my body and made me want to not just be healthier, but be the healthiest I've ever been to create new life. It has given me and my husband James the light of our lives; our little boy Sebby, who's almost five years old at the time of publication.

If this book plays even one tiny part in influencing you to eat better and give your hormones the best chance of being balanced and happy, then I will have accomplished my mission.

The only thing stopping you is you. Do this for you because the greatest investment you will ever make is the investment you make in yourself.

How to Use This Book

I have created targeted recipes using specific, nutrient-dense foods to nourish and support your key body systems. This will help you feel better in those areas, depending on what your body needs from you at any given time. For example, if you are experiencing bloating or constipation, you would flick straight to the gut chapter (see page 65) and follow my high-fibre, gut-feeding recipes to help you regulate your digestive system and feed the bacteria in your microbiome (home to the trillions of microbes in your gut).

Hormones impact every single body system and getting to the root cause of your hormonal imbalances requires a whole-body approach.

I have organised the book in a body-system-focused way because there is a strong connection between every single body system that I cover and how balanced your hormones are. This is an entirely fresh approach.

A WHOLE-BODY APPROACH – AND WHY IT MATTERS

In each chapter, I give you a simple breakdown of how your hormones are impacted by the body system in question and how you can support it to reduce any symptoms you might be experiencing and eat your way to happier, more balanced hormones.

This whole-body approach is how I healed my hormones, how I reversed my PCOS, how I conceived my little boy naturally, how I grew my hair and nails, how I cleared up my skin and how I took control of my mental health. Trust me, it worked, not just for me but for the hundreds of women I've treated since with The Nourish Method.

'RECIPE STACKING'

You can start building yourself weekly meal plans by 'recipe stacking'. This means taking recipes for various body-system chapters and creating a week's worth of meals to support all your needs, from gut to liver to stress management. To make it super easy for you to build your own meal plans, I've included colour-coded icons on each recipe to indicate which other body systems it can support.

Don't forget you can also adapt most of my recipes from any of the body-system chapters to suit the cycle phase you are in by adding in the specific recommended foods in the cycle-syncing guide (using my 'nutrition cheat sheet' on page 218), with the exception of the raw recipes; please avoid eating raw foods during part 2 of your luteal phase (the second half of your menstrual cycle, 5–7 days prior to your period) and during the menstrual phase (the first part of your cycle, when you're on your period) – from an energetic and digestive perspective this matters.

THE NOURISH METHOD

Introducing you to The Nourish Method, a practical and intuitive female nutrition approach, using food as your daily medicine for optimised cycles and balanced hormones.

The Nourish Method encompasses key Eastern and Western food principles and rituals to nurture and balance each body system, aligned to the ebbs and flows of your menstrual cycle. A way of nourishing the whole of you to heal and balance the whole of you, from the first day of your first period to post menopause. Because when you balance your hormones, you balance your life.

The Nourish Method Cycle Sync

If you are following The Nourish Method cycle-syncing guidelines and are eating aligned to the four phases (and seasons) of your menstrual cycle, you will pick foods and recipes from the various body-system chapters accordingly. Please be guided by the seasonal icons in each chapter to indicate specific recipes that align to the phases of your cycle to support optimum hormone balance.

BE CREATIVE, IF YOU WANT TO BE

If you enjoy cooking and are a natural foodie, then I encourage you to tweak recipes according to what you have in stock and what's in season at the time. I want my recipes to be inclusive of all budgets and taste preferences, so change the protein source to what you fancy or like most and please experiment with herbs and spices – this is a great way to keep things interesting.

If you don't enjoy cooking, then keep things super simple. Perhaps start with just one recipe per week and gradually build your own meal plans of your favourite hormone-balancing recipes. Before you know it, you will have a bank of go-to recipes that your body will love you for.

Who is This Book For?

This book for is for all women (anyone who identifies as female) who want to optimise their hormonal health and feed themselves to better health and wellbeing, using food and cycle syncing as medicine to promote healing and restore hormone balance.

I want you to feel empowered by your hormones, replacing dread and negativity with feeling at an advantage – that you can confidently enjoy nourishing your body in alignment with your hormones and cycles. This will change throughout your life, but wherever you are on your hormone journey, you can still eat and treat your body with the love, kindness and respect it needs and will thank you for. The thanks will manifest as increased energy, feeling refreshed and vibrant, creative, motivated, calm and happy – it will feel different in different people, but I can promise it will feel good. Really good.

You can follow The Nourish Method Cycle Sync if you are peri- or post-menopausal, as eating in this way will promote the balanced internal terrain and promote a healthy female body and mind. You can be guided by the seasons (as outlined in each phase of the cycle-syncing guidance in Chapter 3) and use your intuition to tap into how you feel at different times of the month and year.

Be aware that perfection does not exist and with all health-enhancement endeavours life will, of course, happen in between. Our bodies and our lives are not designed to be linear, and this is all part of the beauty of this hormone journey: learning to make peace with the fact that it was never supposed to be perfect and we don't have to have it all figured out. But by the time you get to the end of this book, you will see that by truly honouring your hormones and working with your body you can access something within you that was there all along and use it to improve your wellbeing tenfold.

'Jess is a leader in her field – a gem within our industry.'

Lauren Murdoch-Smith,
former Beauty & Wellness Editor at British Vogue'.

However you identify – whether you are cis-gendered, transgender, non-binary or unsure how you feel at the time of reading – please know that the principles and advice in these pages have working *with* your body at their core, and, as such, they are open to everybody. While I have written this book for women generally and use the words 'women', 'woman' and 'she' and 'her' pronouns, please know that by following my method and recipes, regardless of your identity or if you have a cycle or not, you will experience increased feminine energy, develop a deeper understanding of your body's needs and give yourself the gift of better health.

Throughout this book, I will teach you how to listen to your body's signals, which will not only create a wonderful, renewed connection to yourself, but will help you to make better, more nourishing choices. There's usually some resistance at first, as you are learning something new, but we all know that nothing worth having is easily won, so push through the resistance and be open to exploring new territory. It might just be the one thing that helps everything else fall into place when it comes to enhancing your nutrition. It certainly was for me. Put in the effort at the start to reap the benefits later. You are worth every ounce of effort.

The No Label Diet

This is my personal philosophy and definition of my 'diet'. It describes a way of eating that has regulated my cycle, balanced my hormones and restored my previously complicated relationship with food.

Taking away labels is one of the ways I allowed myself to be truly intuitive with my eating habits and listen to what my body needs and asks for at different points in life. We are evolving beings, what serves us in some seasons of our menstrual cycles and life is not necessarily what will serve us in others and this is ok, it's normal. Fighting against what our bodies ask for is counterproductive and it stops us from being able to truly nourish our bodies and therefore optimise our health. The no label diet means exactly what it says, it's free from labels, which means you don't have to pigeonhole your diet; if you fancy just plants then go for it, if you feel like some fish or meat alongside your veggies then embrace it. Your body has a very clever way of asking for what it needs.

I used to define myself as vegan, and then 'plant-based', and then pescatarian and then felt guilt when I craved meat again post birth, almost like I'd let myself down and worried what people might think of me changing my diet. This sounds silly when writing it down, but it was also such a turning point in my health. It's when I allowed myself to honour my body's true needs, eat the food I wanted without wasting time worrying about how I would describe my eating habits to other people or which 'diet box' it would fit into.

My method of eating is based on intuition and nourishment; and simply eating the food that makes you feel good, which may be different for different people. This works perfectly with The Nourish Method Cycle Syncing guide as I love to adapt the ingredients in my meals to the cycle phase I'm in to go the extra mile for my hormones and to ensure I'm nourishing my body with maximum nutrients to benefit the natural ebbs and flows of my cycle. It's simple, it's an easy way to create more diversity in your diet and supports all of your body systems from your gut to your liver and neuroendocrine system.

Are you ready to join me in the no label diet movement for the ultimate food freedom and joy?

A Note on the Recipes

This is my collection of delicious 'food-is-medicine' hormone-balancing recipes that I love making in my kitchen to nourish my body systems with. Each has been created with hormone balance at the forefront, but the ingredients – carefully chosen to support your key body systems – will help to heal the whole of you for optimised hormonal health and wellbeing.

Digestion starts with your eyes . . . the more vibrant colours there are on your plate, the more signals your brain will send to your gut to prepare to digest, and the more nourished your body will be.

A healthy diet is more than just macro- and micronutrients. It's a sensory experience that should spark joy and happiness and make you feel good emotionally, as well as physically nourished. Enjoy the food you eat, enjoy the daily ritual of supporting your body and remind yourself that making healthy choices around what to eat is not a chore – it's actually a privilege that you are lucky enough to enjoy each and every day, and it means you will never, ever need to diet again. **Health is wealth, and eating healthily is a way of life that your hormones will thank you for now and keep on thanking you for in the future.**

MY COOK-ONCE RULES FOR FAMILY COOKING

If I've learned anything about children's nutrition and eating habits since becoming a mum of a hungry little boy with a bottomless pit for a tummy, it's that getting them involved in the kitchen gives them a sense of achievement and encourages them to want to eat the food they have made. Plus, it's the perfect bonding time.

Talking positively about all types of foods and plants, what they look, feel and smell like, and where they actually come from gives them more context and makes them matter more. I receive lots of questions on social media about Sebby's diet, asking how I get him to eat vegetables and to eat what I eat. The truth is, I don't stress when he doesn't eat everything on his plate. I trust his intuition and the next mealtime is a new opportunity for him to nourish his body using the same approach I have to my own diet. I might have a pizza and a glass of wine on the odd occasion, but the next day is a new opportunity to renourish my body.

I love to make Sebby variations of the same meals we eat and prioritise eating together as a family when possible. This makes a profound difference when it comes to how excited he is by the food on his plate and the enjoyment he gets from mealtimes – which, of course, results in him actually eating it. Mission accomplished!

Many of the recipes in this book can be easily adapted for children's plates by sticking to the core of the meal (the main source of protein and carbohydrates in the recipe), but adding in a few favourites you know they like, such as berries, cucumber, avocado or pepper fingers, some pasta or homemade sweet-potato 'chips'; this helps to make it more familiar and means you can eat together and only cook once – every parent's goal, right?

TIP: in the recipes you can swap out the green herbs for any other herb you fancy.

1. Eating Intuitively For Hormone Balance

'Your body hears everything your mind says.'

Naomi Judd

I talk a lot with my clients in clinic about the endless benefits of 'tapping into our intuition' and how we can use this internal superpower to our advantage to support nutrition and improve our relationship with food without ever having to restrict, count calories or diet again. It sounds too good to be true, but intuitive eating is a real thing. It works, and your hormones will benefit from it, I promise.

What is Intuitive Eating?

Intuitive eating is learning to eat intentionally and mindfully by tuning into your body and listening to the signs and signals it gives you. It should, in fact, be simply called 'eating' because it's what we are instinctively designed to do.

Children are masters of intuitive eating. You will notice they just do it – it's not something they are aware of or try to do. If you stop and pay attention to their eating habits, you'll see that they naturally stop eating when they're full and ask for food when they are hungry.

We should be able to do this naturally, too, and we can easily reignite our intuition around food. It's within all of us and our bodies are desperate for us to use this health-enhancing tool we are born with.

Ultra-processed Diets and the Death of Intuitive Eating

Because the standard Western diet is so high in UPFs (not just chips and burgers but also cereals, sauces and other foods that we've been told are 'healthy' but are far from it) and laden with chemicals we can't even pronounce, our instincts around food have been dramatically altered.

This is because UPFs stimulate our hunger and satiety hormones, ghrelin and leptin.[1] Ghrelin, known as our 'hunger hormone', is produced in the gut. It's secreted into the bloodstream when our stomachs are empty and communicates with a part of the brain called the hypothalamus, which helps to regulate hormones, appetite and weight.[2] This is all well and good when we are eating healthy wholefoods as nature intended, as we then know we can rely on the accuracy of the signal that ghrelin sends to the hypothalamus to warn us that we need to eat and we should act accordingly. But frustratingly, chemicals in UPFs can stop this signalling process from happening and result in us overeating, causing us to overburden our bodies, leading to impaired health and links with chronic health conditions.[3] In fact, multiple studies confirm that a diet high in UPFs not only causes overeating and weight gain, due to dysregulated ghrelin signalling, but is also associated with hormonal conditions such as PCOS and insulin resistance (when your cells don't respond properly to the insulin your body makes).[4]

Leptin, on the other hand, is our satiety hormone.[5] It's produced by fat cells (adipocytes) and plays a key role in regulating energy balance and appetite, as it acts on receptors in the hypothalamus to suppress appetite, decrease food intake and increase energy. Its primary function

is to signal when we've eaten enough when energy stores are sufficient. This signalling process happens when fat cells are filled with stored energy (triglycerides) and release leptin into the bloodstream, preventing us from eating more than our bodies need.

Diet culture and the miscommunication and marketing of processed foods as 'health foods' are a big problem when it comes to the imbalance of these key hormones that dictate appetite and influence our weight and hormonal health. However, by minimising your intake of chemical-laden foods and increasing the number of real wholefoods you eat, you will be actively supporting your body (and maintaining a healthy weight as a result).

It's Not Your Willpower, It's the Chemicals in Your Food

If you've previously thought the reason why you haven't been able to stick to eating healthily was a lack of willpower, I hope I've shown you that it's less about that and much more about the types of food you put into your body. It's the pro-inflammatory chemicals in foods that make it challenging for you to eat intuitively because of their impact on your hunger-hormone signalling.

TIP: Eat until you are 80 per cent full to allow your brain the twenty minutes it needs to catch up with your gut and receive the signal that you are full (from your hormone leptin). This supports digestion and prevents overeating.[6]

TOP TIPS FOR EATING INTUITIVELY

Tune in before, during and after your meal.
First things first, how hungry are you? Honour your body's *true* needs. Don't deprive, starve or overburden it – if you do, it will let you know about it later (if you're listening clearly enough). Tune into your genuine hunger cues without judgement; even if hunger presents itself outside of your three meals, don't starve yourself – this is a disservice to your hormones. If you really tune in and listen to your body's satiety cues, leptin will signal when you are getting full, and by listening to this signal instead of ignoring it you will be putting intuitive eating into practice. It can also serve as a helpful feedback loop. For example, if you find you are regularly hungry in between your main meals, then you likely need to look at how you are building your meals and if you are eating enough complete protein (one that contains all the essential amino acids needed for optimal protein synthesis, energy production, immune function and nutrient absorption) from chicken, fish, eggs, dairy, quinoa, buckwheat and soy products. Including a complete protein and ensuring half your plate is filled with brightly coloured vegetables and fibre will keep you sustained and promote balanced blood-sugar levels.

How does your food look and smell?
Embrace the tastes and textures of what you are feeding your body, notice the variety of colours on your plate, take in the smells of the food you have spent time preparing and feel the different textures in your mouth as you enjoy your meal. Remember, the more colour you can see in your meals (from natural wholefoods), the more hormone health benefits you will be gifting your body with.

Eat slowly, sitting down at a table and away from the distraction of screens.
Do you eat on autopilot and rush through your meal? Is this a habit? If so, place your knife and fork down on the table between each mouthful to help physically slow yourself down and break the cycle of eating too quickly. Eating slowly while sitting down, away from the distraction of screens, will aid digestion, reduce the chances of overeating and help you to connect with your body, allowing you to become more mindful at mealtimes and increasing the joy and overall experience of healing eating.[7] This is because the brain finds it challenging to signal to your gut that you are starting to get full when it is focusing on other things (screens, for example), and will therefore increase the speed at which you eat, leading to weight gain, according to research.[8, 9] Eating more slowly will keep you feeling fuller for longer.

Chew your food (very) well.
Chewing your food well will stimulate your digestive juices, helping to improve digestion (because this process starts in the mouth) and increase absorption of nutrients. The term 'you are what you absorb' is so relevant here – the more nutrients you can supply your hormones with, the more balanced they will be.

This simple chewing hack will also enhance your gut–brain axis (the physical and chemical connections between gut and brain) and will improve how accurately your brain recognises fullness cues to prevent overeating.[10] Multiple studies show that increasing the number of chews per bite of food increases levels of gut hormones that make you feel full.[11] Aim for the food in your mouth to be as mushy as possible (roughly twenty to thirty-two chews per mouthful is thought to be optimal).

Recognise and honour the sensation of fullness.
Acknowledge the true sensations of hunger and fullness within your body and specifically your gut (your second brain) to avoid feeling full to the point of discomfort.[12] This will prevent you eating more than you need by helping you to stop before you get to this point. You will become familiar with knowing exactly when you have eaten enough to feel good and comfortable, and the same applies to eating enough to feel fully nourished and content. Undereating is as damaging to your hormonal balance as overeating.

Notice how the food you've just eaten has made you feel.
Physically, do you feel heavy, sluggish or too full? Emotionally, did this meal spark joy and happiness? Food has the power to energise or deplete you. It's your choice what you put into your body, so don't forget your health is in your hands.

Take four deep breaths at the start and end of your meal to stimulate your parasympathetic nervous system.
A calm and relaxed body means a happy body and mind. Breathing is a superpower and an important part of learning to eat intuitively. Taking four breaths at the start and end of a meal is a free resource we all have access to but one we often forget to utilise. Calming your nervous system is going to make it a lot easier to tune into your senses and put these tips into practice.

These tips will help you to actively tune into your senses and receive maximum joy from your mealtimes, instead of just eating on autopilot and the whole thing being over almost as soon as it started. Take a few minutes to soak up your mealtimes, as they are moments of self-care in your day, each being a new opportunity to support your health and improve your hormones.

If you are serious about becoming more tuned into your body and working with it for the greater good of your hormonal health (which I promise will make a positive impact), then trust me – it only takes a few minutes to check in with yourself, notice what's going on internally and to eat according to the signals your body is giving you. Eating intuitively is like training a muscle – the more you do it the stronger those skills will become, until they are simply your normal.

COUNT COLOURS, NOT CALORIES

If you are struggling with restriction around calories, then this is for you.

Calories measure the amount of energy a food provides, but the food you eat is so much more than that. Calorie counting is outdated and totally irrelevant to hormonal health because it uses a system of averages, and guess what? We are not average! The calories in food are also altered when heated as opposed to uncooked and are influenced by how well (or not) you absorb food.

But the biggest issue I have is that the calories printed on a food packaging do not take nutritional value into account. The message around calorie counting for healthy diets is so misleading because the calories in nutrient-dense wholefoods may be higher than those in ultra-processed foods (UPFs). But this small piece of information totally ignores any preservatives, additives and other hormone-depleting chemicals found in UPFs, so food choices based on calories alone are not made with good health in mind. For example, avocados, olive oil, nuts and seeds contain fats that our hormones require for production, as well as fibre for metabolism and elimination and are loaded with hormone-balancing vitamins and minerals like selenium, magnesium, vitamin E and zinc to support everything from regular PMS-free cycles to a healthy thyroid, clear skin and optimal fertility.

I urge you to start putting nutritional value at the forefront and step away from counting calories, instead focusing on the colour and variety in real foods you can see on your plate. This is a fundamental part of my method and an approach that will benefit every single system in your body. Counting colours from plants and wholefoods on your plate (rather than meaningless numbers, i.e. calories) will instantly provide a much wider range of nutrients, and will make you healthier, happier and hormonally balanced.

Make a Commitment to Your Hormones

Now that you know how to start eating intuitively, I propose the first step in your hormone-balancing journey should be a commitment to eating healthily by following the carefully curated meal plans at the back of this book (see pages 220–223) or, at the very least, simply committing to limiting UPFs and focusing on eating fresh, colourful wholefoods at each meal. Research carried out by Professor Tim Spector and his team shows that eating thirty plant varieties each week is associated with a healthier gut microbiome which, in turn, results in a healthier body overall and a reduction in hormonally led conditions.[13,14]

This is going to help you hear what your hunger hormones are telling you, which will, in turn, help you to become increasingly intuitive with your eating habits, so that you can gravitate towards healthier foods (and portions) that work for you and your individual needs. Eating intuitively is perfectly aligned to cycle syncing, which we're going to dive into in Chapter 3.

2. Getting To Know Your Hormones and the Reproductive System

'Look after your hormones and they will look after you.'

With female health conditions getting just 1 per cent of global research funding, there couldn't be a more significant time to be your own hormonal health advocate than now.

Your reproductive system comprises your reproductive organs, including your vagina, uterus (womb), ovaries and fallopian tubes. This special system also creates and houses the hormones responsible for your fertility, your period and your sex drive. It may not be the heart of your body, but it's the heart of your hormonal health, so it makes sense to start here.

I'm not going to give you a science lesson in the biology of your body – there are already plenty of resources out there for that.

But I am going to invite you to look at your reproductive system with fresh eyes with the aim of guiding you to nourish it like it's never been nourished before.

My view is that up until now the female reproductive system has had most of its airtime in the context of either trying to prevent or encourage a pregnancy. And yes, I know it's called our 'reproductive' system for a reason, but it's so much more than that and this is largely forgotten until something malfunctions, at which point we give it the attention that it has, in fact, required all along in order to impact us in so many positive ways beyond regular cycles.

Working With Your Reproductive System, Not Against It

Instead of celebrating the start of our menstruating lives and being taught to embrace our monthly bleed and use it as a time to rest and recuperate, we usually end up dreading 'the time of the month' – whether consciously or otherwise – due to a culture of shame and embarrassment around the female body. Growing up, among my group of girlfriends, talking about periods was a negative thing. We would whisper the 'P' word in front of the males in our lives, hide it and almost be ashamed of it. And this was standard procedure. In fact, I don't know any families who actively, openly talked about periods in a positive way. Nor do I recall it ever being talked about as a positive part of womanhood on TV, in books or blogs. And I absolutely know that any reference to the different phases of the menstrual cycle was unheard of. I wasn't even aware there were different phases, or that adapting my nutrition had the power to stop the horrendous PMS I used to suffer with as a teen, whose period started at almost seventeen years of age, long after all my friends.

The Nourish Method is about giving you a fresh perspective, offering a new lens through which to see, understand and optimise your cycle. To liberate you, so you can stop ignoring it or labelling it a liability, an inconvenience or even a burden.

This extends to our reproductive health and our cycles. So many women find that instead of getting to the root cause, our GPs (of course not always, but from experience, sadly, more often than not) somewhat fail us by being too quick to prescribe the oral contraceptive Pill (one of the biggest regrets of my hormonal life) without explaining the effects that this will have on our hormones and, therefore, our bodies as a whole. For me, this happened at an age when my hormones had yet to establish their own rhythm, and caused many issues, until, years later, I was able to educate myself and take back control. And I know I am nowhere near alone in this.

Yes, things have changed for the better in recent years, with far more mainstream conversations and education happening around women's hormones, which is an incredible step forward, but there's still a long way to go. I still have a clinic full of women who are beside themselves, with PMS, non-existent cycles and struggling to conceive.

Surely if we continue to ignore and go against our hormones (as a result of not proactively supporting them), it should be no surprise when, one day, they wear down and stop playing ball? Can you blame them?

HORMONE LEVELS

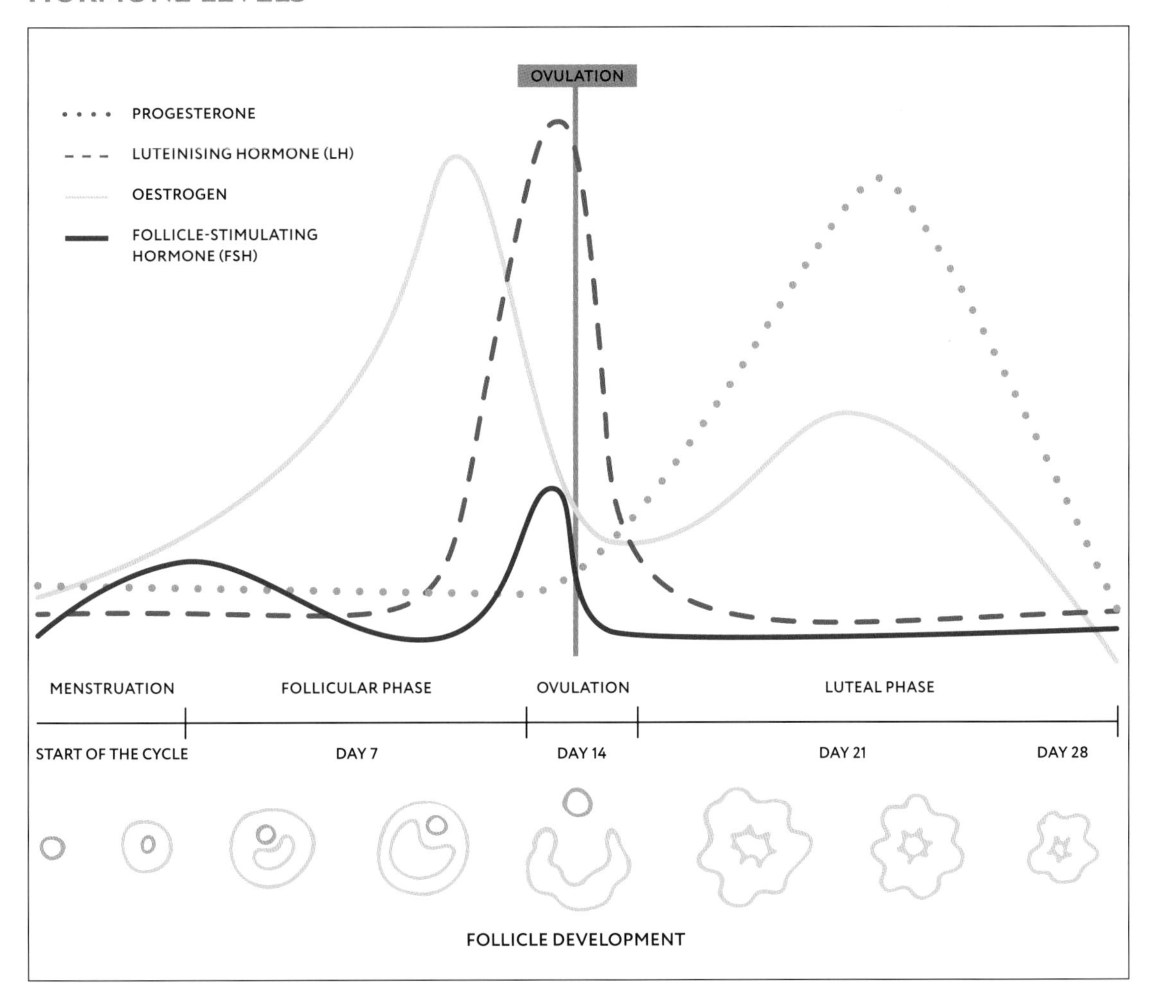

What Are Hormones?

We've talked a lot about hormones so far, but what actually are they?

Your hormones control pretty much everything in the body, including growth and development, influencing every cell, organ and function. They are chemical messengers produced in your endocrine glands and secreted in your bloodstream, where they travel around, telling your tissues and organs what to do, playing an integral role in many of your body's major processes. The degree to which your hormones peak and plummet over your monthly menstrual cycle plays such a crucial role in your body's responses and massively impacts your emotions, appetite, weight, libido and mental health, too (hello, luteal-phase anxiety anyone?).[15]

Most of us are aware of the role of hormones in sex and menstruation, but who would have thought they also have the power to impact gut health and digestion, liver detox, how your hair grows, how clear or glowing your skin is and how anxious you feel. It's quite astonishing just how big a role our hormones play in most of the key inner workings of both our bodies and minds. And that is exactly why I've dedicated a whole book to helping you nourish them to better health via the food you eat and *when* you eat it to help optimise each major body system.

YOUR REPRODUCTIVE HORMONES

There are seven key hormones we focus on throughout the reproductive system and cycle-syncing chapters, as follows:

Oestrogen

An umbrella term covering oestradiol, oestrone and oestriol, oestrogen (as a collective) is the queen bee of your cycle; she likes to be the star of the show and, from a traditional Chinese medicine (TCM) perspective, is your 'yin' hormone. She's an expert at making you feel confident by boosting your neurotransmitters serotonin (your 'happy' hormone) and dopamine (giving you motivation and pleasure). Oestradiol is the type we will be referring to the most throughout this book – it's released by your developing follicles and dominates the first half of your cycle (the follicular phase). It is responsible for stimulating the thickening of your uterine lining in anticipation of a pregnancy (regardless of whether or not this is your intention). It also stimulates your 'fertile mucus', aka ovulation. Oestradiol is a key player in lots of other vital biological systems, protecting you from bone-density loss (osteoporosis), cardiovascular disease and dementia. You also have oestrogen to thank for your skin looking plumper and clearer during your follicular and ovulatory phases and for a boost in energy and confidence.

Progesterone

The yang to oestrogen's yin, this counterbalances oestrogen and is your calming hormone, as it likes to relax your nervous system by promoting sleep, regulating mood and helping you deal with stress. The main responsibility of progesterone is to hold and maintain a pregnancy ('pro-gestation'), but it does a lot more than this. While oestrogen thickens the uterine lining, progesterone counterbalances this by thinning it, and while progesterone boosts thyroid hormone, oestrogen supresses it, increasing our need for thyroid-stimulating hormone (TSH). This effect on your thyroid causes a rise in body temperature in your luteal phase (helping you to determine when you have ovulated). Progesterone production happens around ovulation, and if an egg doesn't fertilise, then progesterone levels drop and the lining automatically sheds (aka your period, day 1 of your menstrual phase). If, however, you don't ovulate, you can easily see how this impacts your progesterone levels (there are many accompanying symptoms) and how this creates an imbalance.

Testosterone

Produced in the ovaries and adrenal glands, it helps you build muscle and burn fat, and it supports energy levels. It's connected to healthy libido, as it helps you connect to your sexuality, which makes it significant when it comes to fertility. Testosterone communicates with oestrogen and cortisol.

Follicle-stimulating hormone (FSH)

Produced by the pituitary gland, it plays an integral role in our cycles because it causes an egg to mature in the ovary and stimulates the ovaries to release oestrogen.

Luteinising hormone (LH)

Again, produced in the pituitary gland at ovulation, it stimulates the release of progesterone. LH is responsible for causing the mature egg to be released by the ovary. Dysregulated levels of LH are often associated with fertility issues and PCOS.

Insulin

This regulates your blood-sugar levels, is produced in your pancreas and is secreted after you eat carbohydrates. Your body breaks these carbs down, turning them into glucose (sugar). This is then absorbed into the bloodstream and your cells use it as energy in your body. The more carbs you eat, the more insulin your pancreas is secreting into your blood, which can be a huge stressor on your body and is often the driver of many hormonal imbalances, including irregular cycles. Regulated insulin levels (aka balanced blood-sugar levels) are key to balanced hormones and optimised cycles.

Cortisol

The release of cortisol is determined by your hypothalamic-pituitary-adrenal axis (HPA). We need a small dose of cortisol for our 'get up and go' and productivity, as well as other functions, but too much or too little can cause issues for your delicate hormone balance. Chronically elevated levels of cortisol can inhibit ovulation, decrease progesterone production and steal your sex drive, all of which cause issues for fertility.[16] Chronic stress/chronic elevated cortisol negatively impacts the feedback mechanisms that help maintain overall hormonal homeostasis.

What Does 'Hormone Imbalance' *Really* Mean?

A 'hormone imbalance', put simply, means having too much or too little of a hormone. It's a delicate balancing act that our bodies are pros at. But sometimes, even a small imbalance left untreated over time can cause issues. Our hormone balance is directly influenced by the food we eat, the exercise we do (or don't do), how much restorative sleep we get, our stress levels and our environmental toxin exposure. Essentially, this means that any of these factors can throw your hormones out of that delicate balance with the potential that it can become the underlying issue for several common health concerns, including irregular periods, PCOS, thyroid disorders, acne and chronic fatigue.[17]

Your internal orchestra of hormones

I love the idea of our hormones (aka our endocrine system – a complex network of glands and organs that uses hormones to control and coordinate our bodily functions) being like an orchestra playing a symphony. When each instrument plays in tune, the result is beautiful, flowing music, but when one is out, it creates chaos. In the same way, when one of our hormones is out of whack, it has a knock-on effect on the others. Sometimes this may be only a minor change that we don't always notice, but over time, it can cause or drive further imbalances, impacting how our bodies function.

Why you should consider coming off the Pill

Hormonal birth control is designed to suppress ovulation (the release of an egg from your ovary). It makes it impossible to tune into what your hormones are telling you because they have been temporarily turned off and you simply can't balance hormones that are asleep. This can be a problem if hormone balance is the goal, because you need to know how you can help and nourish your hormones to be at optimum health.

I feel strongly that you don't need to shut down your entire hormonal system just to prevent a pregnancy. Thankfully, there are many hormone-free alternatives, such as tracking your cycle and measuring your basal temperature using the Natural Cycles app, Oura Ring, the copper IUD and male condoms to help prevent unwanted pregnancy, while leaving your hormones well alone.

From my personal and clinical experience, the Pill often puts a plaster over hormonal imbalance issues. For example, if you were prescribed the Pill for your skin, you should remind yourself that yes, while it is working for you now because it stops skin-oil production and dries up your spots, it only works for as long as you take it. So when coming off the Pill, I urge you to work alongside a naturopathic nutritionist who can help naturally bring your hormones into balance and minimise associated negative symptoms.

If you went on the Pill to regulate your period or to manage your PCOS, you cannot regulate a period that's been switched off because a Pill bleed is not a period, it's a Pill-withdrawal bleed. A real period reflects your hormones and overall health, it's your monthly health report card.

Something else to consider, when you are on the Pill is that your body is running off synthetic steroid hormones instead of your real sex hormones (the ones that are in a deep sleep) and these are known to deplete essential nutrient stores needed for optimal health and wellbeing, including folic acid, vitamins B2, B6, B12, C and E and the minerals magnesium, selenium and zinc.[18]

One of my biggest regrets is that I ever went on the Pill in the first place. I know it was at the root of my hormonal issues, including my dysregulated thyroid, and it makes me upset that it's so freely given to girls as young as fourteen, whose hormones haven't even had a chance to regulate in the first place. If you still feel the Pill is the best option for you and your lifestyle to prevent pregnancy, then that's great, you do you, but it's important you know the facts and future potential considerations, as these are often not explained.

DIM (AKA DIINDOLYLMETHANE)

This is a natural compound created after digesting vegetables that contain the chemical indole-3-carbinol, including broccoli, Brussels sprouts, cauliflower, and cabbage.[19,20] It helps to metabolise oestrogen effectively, which is particularly useful in cases of oestrogen dominance. Studies show you would need to eat high amounts of these vegetables to get the full benefits, so supplementation is recommended for cases of excess oestrogen (consult with a healthcare professional for tailored guidance).

OESTROGEN EXCESS

A common example of hormone imbalance and one I see in clinic daily is when there's too much circulating oestrogen (this means high oestrogen compared to progesterone). The relationship between these two hormones is crucial. Signs of excess oestrogen include heavy menstrual bleeding, breast tenderness, PMS, painful periods, menstrual migraines, endometriosis, fibroids, cysts, mood swings, weight gain around the middle, brain fog, water retention, abnormal smear-test results.

Oestrogen imbalance can be caused by:

- anovulatory cycles (cycles where ovulation did not occur and therefore progesterone was not made)
- PCOS
- impaired oestrogen detoxification
- poor diet
- histamine intolerance
- gut dysbiosis (when the bacteria in your gastrointestinal tract – which includes your stomach and intestines – become unbalanced)
- constipation (as oestrogen is reabsorbed)
- elevated cortisol (this blocks progesterone receptors)
- heavy alcohol consumption
- diabetes
- some autoimmune conditions.

To maintain overall hormone balance you should:

- support liver detox
- support the gut microbiome by increasing fibre intake to promote oestrogen detox via your poo
- increase hydration to promote oestrogen detox via your wee
- prioritise sleep
- reduce or eliminate alcohol intake
- exercise regularly to promote oestrogen detox via your sweat
- lose excess weight
- eat cruciferous vegetables including broccoli, cabbage and Brussels sprouts, which contain compounds called

indole-3-carbinol and diindolylmethane and may help to metabolise oestrogen in the body and reduce excess levels (except in the menstrual phase, when raw cruciferous veg should be avoided).

> **TIP:** Broccoli sprouts are known to be beneficial for the clearance of oestrogen by the liver, which enhances sulfation, a liver detoxification pathway, which helps, in turn, with the clearance of oestrogen by the liver. You can easily sprout them yourself at home, which makes them an affordable way to support your hormone balance.

PROGESTERONE – THE CALMING HORMONE

Low progesterone can cause PMS, anxiety, depression and worsen perimenopause symptoms. Fasting and cutting carbs are two of the worst things you can do if your progesterone is low because undernourishing elevates your cortisol, and cortisol hinders progesterone production.

Eating complex carbs and increasing magnesium intake from nuts, seeds and leafy greens and supplementation during the luteal phase of your cycle are two positive things you can do to encourage progesterone production. This soothes your nervous system and provides your body with the glucose it needs for healthy ovulation. Signs of low or less-than-optimal progesterone include:

- no luteal phase or short luteal phase
- fertile mucus during the premenstrual phase
- PMS
- premenstrual bleeding or spotting
- prolonged or heavy menstrual bleeding
- acne
- hair loss
- and low progesterone is associated with PCOS, fibroids and perimenopause.

Why Ovulation Is Not Just Important, But Essential

When it comes to healthy cycles, ovulation is everything. It is usually only given airtime when we want to fall pregnant, but it should be the goal of every woman who wants to balance her hormones.

Ovulation is how you make the incredible hormone progesterone which, as we've seen, is essential for healthy periods, mood and metabolism. If you don't ovulate, you haven't produced progesterone, and if you haven't produced progesterone, then this impacts oestrogen, and you will be hormonally imbalanced. It's as simple as that.

To determine whether you have ovulated or not, you can take your basal body temperature (your waking temperature), which will rise when ovulation occurs. This indicates the release of progesterone (measured by a mid-luteal-phase blood test) and it will continue at that higher level until your period comes.

But just because you have a monthly period, that does not mean you have ovulated. It is important you pay attention to this for the health of your hormonal system.

Physical signs of ovulation include fertile cervical mucus (aka discharge) that looks and feels like raw egg white. It's clear, stretchy and slippery in appearance and you will notice it on the toilet paper as you wipe and in your underwear in the days leading up to ovulation, but you can also see it early in your cycle if you have high oestrogen levels, too, and after ovulation if you don't make enough progesterone. So while you can only ovulate once in your cycle, you can see mucus multiple times.

What Happens to Hormones During Perimenopause?

Think of perimenopause as the reverse of puberty. Perimenopause is the time when your body's reproductive system slows down, until you hit 'menopause', the official date marking the full year since your last period.

During perimenopause your ovaries prepare to stop releasing eggs entirely by producing less oestrogen, progesterone and testosterone, and with this wind-down comes unpredictable and changeable cycle behaviour. Some months you will ovulate, while in others, no egg will be released. The hormonal fluctuations can become quite erratic and feel chaotic, when one minute you will feel like you and the next, you might not recognise yourself. It can be a rollercoaster.

WHAT'S THE SCIENCE SAYING?

Follicle-stimulating hormone (FSH) levels rise and fall, causing oestrogen to do the same, while progesterone stays low as you begin to ovulate less, and your cycle becomes more irregular. Unsurprisingly, these fluctuations have a knock-on effect on the whole hormone system, particularly oestrogen, as there are oestrogen receptors throughout the body.

The best possible advice is to be your own hormonal-health investigator, just as you would if you were going through any hormonal shift; whether you were in your twenties and thirties, from coming off the Pill to preparing for conception, your hormones need to know you have their backs, and that will change the course of your journey through perimenopause to menopause and beyond.

Track your symptoms and changeable cycle by noting down exactly when your mood, energy and mental health change, along with physical changes, and carry out a nutrition and lifestyle audit (as I like to call it when talking to my clients about this) on what might be amplifying or driving your symptoms. Is there anything you eat or do that has crept into your day-to-day – are a lack of protein, or too much sugar or alcohol and late nights making your symptoms worse? Use this book to work with your body and you will feel the benefits.

When You Balance Your Hormones, You Balance Your Life

Eating with intention and respect for your hormones' inevitable fluctuations will help you naturally bring them back into balance, while encouraging you to step into your feminine energy and harness its power, so that you feel and show up in the world as the best version of you. The more you feed and treat your body to balance what your hormones are doing internally, the more your reproductive system will pay you back with balance, strength, peace and vitality.

You want to be sending a message to your body that there's plenty of nutrition, so as to prevent it from downregulating reproduction. By restricting food, you restrict the building blocks your hormones need to create a healthy monthly cycle and to create new life.

In the next chapter, I'm going to share practical ways to help your internal orchestra (your hormones) play its symphony, just as nature intended by introducing you to my cycle-syncing method. This will show you how to eat to support your reproductive system in alignment with the four phases of your cycle to optimise hormonal health. This is important because each phase impacts how we feel, what we crave and what we need to eat in order to support these changing phases.

3. Your Cycle-Syncing Guide

'Let thy food be thy medicine and medicine be thy food.'

Hippocrates

The cycle-syncing guide in this chapter looks at what your hormones are doing at each phase of your cycle and how to nourish them into balance.

Menstrual Phase

Winter

Days 1–6 of your cycle

Duration: 3–7 days

Priority: Restoration and Micronutrient Replenishment

Your menstrual phase is an invitation from your body to slow down and retreat inwards. It's the ultimate time to level up comfort, warmth and mineral-rich nourishment as your hormones are at their lowest point in your cycle and your body is busy shedding the built-up lining of your uterus.

Focus on increasing iron- and zinc-rich foods to replenish lost nutrients through blood loss and restore the body's micronutrient stores, such as grass-fed meats, legumes/lentils, dark, leafy greens and sea vegetables. Serve with a source of vitamin C to increase absorption. Magnesium will help reduce menstrual cramping as it acts as a muscle relaxant. Easy food sources include leafy greens, nuts, avocado and dark chocolate (the darker the better) and dark-coloured berries will provide your body with a dose of much-needed antioxidants.

Give your body extra targeted TLC by prioritising an increase of complete proteins and healthy fats (especially omega-3 fatty acids). The amino acids from the protein will support hormone synthesis as well as stabilise blood sugar levels, and studies show that healthy fat intake will help set you up later for healthy ovulation, and encourage progesterone production.

You are in the coldest phase of your cycle, your inner winter, according to TCM. This is when your body temperature has dropped, so avoid cold, raw foods, especially uncooked cruciferous vegetables. Your body will benefit from cooked, warming, hearty and wholesome meals that will heat you up from the inside out and energise the cells in your body.

Eating warming cooked foods will also support your digestive system, especially as it's more sensitive during this phase of your cycle. So opt for meals that are easier to digest, such as stews, soups, broths, curries and slow-cooked meals packed with anti-inflammatory compounds including ginger, garlic and rosemary (which will also reduce menstrual cramping).

Avoid restricting and fasting, because getting enough nutrition at each meal to feel satisfied will give your body what it needs to naturally help increase your energy levels and elevate your mood by supporting your neurotransmitters serotonin, dopamine and GABA. (Think of GABA as your 'anti-anxiety' neurotransmitter that stimulates a feeling of inner calm. It's a naturally occurring amino acid that works in your brain to regulate mood.)[21]

Drink warming, anti-inflammatory tea, such as chamomile, an anti-spasmodic to ease the tension in the uterine muscles, providing relief from cramping. And choose room-temperature water over cold.

Follicular Phase

Spring

Days 7–12 of your cycle

Duration: 7–10 days

Priority: Supporting Metabolism, Growth and Renewed Energy

The levels of oestrogen and luteinizing hormone (LH) begin to rise after winter, melting the frost and bringing a lift in mood, motivation and a sense of renewed energy and growth, making us feel more like ourselves compared to the menstrual phase.

This, the spring of your cycle, is a great time to maintain some warming nourishment, whilst introducing fresh, fibre-rich, crunchy foods to support liver function and hormone detoxification post period. Increase the quantity as you move towards ovulation.

Introduce phytoestrogens, such as flax and (organic) edamame, as they contain plant compounds that mimic the body's natural oestrogen, helping to balance you. It starts to build up in preparation for ovulation. Adding in choline-rich foods, such as eggs, will support the quality of the eggs maturing in your ovaries.

Nourish your microbiome by adding in pre- and probiotic-rich foods like kimchi and sauerkraut to support estrobolome (a collection of bacteria in the gut, capable of metabolising and modulating the body's circulating oestrogen) for healthy hormone metabolism.[22]

The foods you eat during the follicular phase will prime your body, ready for healthy ovulation and progesterone production, and will help to minimise PMS during the luteal phase.

Ovulatory Phase

Summer

Days 13–15 of your cycle

Duration: 3–4 days

Priority: Elimination and Detoxification

As you transition from the spring of your cycle to your inner summer, and as oestrogen and LH peaks, you should feel a surge of energy, a stable mood, increased sex drive and a zest for life. Bonus – your skin and hair is likely to be its glowiest too!

This is the phase to embrace lighter meals because your metabolism is slower in the first half of your cycle. You will feel satisfied more quickly with lighter grains (and fewer carbs) than you do in the other phases, as your body requires less of an energy boost.

This being the hot phase of your cycle, from a Chinese-medicine perspective, and your digestive system being at its strongest, you can handle most raw foods, making it an ideal time to embrace fresh salads and raw cruciferous vegetables for their cooling effect on the body. Their high sulphur content boosts your glutathione levels, helping your liver to metabolise oestrogen from your body more efficiently to balance levels and prevent an excess, and any associated symptoms.

Prioritise fibre-rich plants and maximise plant diversity in this phase, to keep your bowels moving, flush oestrogen that your liver is metabolising and support overall hormone balance. Constipation can cause a recirculation of oestrogen, worsening PMS, bloating and acne. Remember to stay hydrated to receive the full benefits of the high-fibre plants you eat, and to help transport the fibre along your digestive tract, so everything is moving in the right direction for hormone balance.

Luteal Phase

Autumn

Days 16–28 of your cycle

Duration: 10–14 days

Priority: Stabilising Blood-sugar Levels

After ovulation, progesterone is released and remains high until menstruation, whereas LH and oestrogen drop (although oestrogen rises again mid-luteal). The second part of your luteal phase is a time to be extra gentle with yourself and to ramp-up the nourishment, because the more consistently you nourish your body the better you will feel.

Your metabolism is faster during the luteal and menstrual phases, as your body has an increased demand for energy and nutrients, such as B vitamins and magnesium (from organic poultry, wholegrains, eggs, avocado and seeds) to support progesterone production and prep for your period.

You not only tend to crave more during this time with increased appetite, but your body actually needs more nutrition to help your reproductive system carry out its job efficiently and to prevent dipping energy, anxiety and low mood.

To support fluctuating hormones and increased resting metabolic rate, include slow-burning complex carbohydrates such as root veggies – sweet potatoes and squash (which help to satisfy sweet cravings, too), high-fibre beans/legumes alongside fats, greens and proteins. This will provide your body with the energy boost it's asking for. Add tryptophan-rich foods such as organic chicken and wild salmon to support mood and sleep.

The slight extra intake of complex carbs will also help to prevent unhelpful sugar cravings and keep hunger at bay by supporting balanced blood-sugar levels from meal to meal. This will encourage you to feel steadier and prevent pesky PMS symptoms – the ultimate luteal-phase goal!

The Nourish Method Cycle Syncing Guide

	MENSTRUAL PHASE	FOLLICULAR PHASE	OVULATORY PHASE	LUTEAL PHASE
VEGETABLES	• Kale, cavolo nero, chard • Beetroot • Mushrooms • Root (such as sweet potato, squash and pumpkin) • Sea vegetables (dulse, kombu, kelp, seaweed) • Ginger and garlic	• Artichokes • Broccoli, kale, lettuce, rocket, watercress, • Green peas, courgettes, carrots • Rhubarb • kimchi and sauerkraut • Sprouted beans • Olives	• Cruciferous: broccoli, kale, cauliflower, collard greens, cabbage, pak choy, Brussels sprouts, radishes, chard, chicory • Spinach • Spring onions • Red peppers • Tomatoes • Celery • Cucumber • Aubergine	• Root: sweet potatoes, squash, pumpkin, parsnips, red onions • Cabbage, cauliflower, pak choy, watercress, spinach • Leeks • Spring greens • Garlic, ginger, turmeric
COOKING GUIDANCE/ FOOD ENERGETICS	100% cooked	60% cooked, 40% (increasing raw as ovulatory phase approaches)	80% raw, 20% cooked	First part 50% raw, 50% cooked; second part 100% cooked
FRUITS	• Dark-coloured berries (blueberries, blackberries, cranberries) • Red grapes • Avocados	• Avocados • Pomegranates • Citrus fruits • Cherries • Plums • Bananas • Peach	• Strawberries, raspberries • Watermelon • Apricots • Coconut • Citrus fruits	• Avocado • Pears • Apples • Dates • Bananas
PROTEIN	• Grass-fed red meat • Chicken • Eggs • Sardines • Mussels • Oysters • Legumes/lentils (such as kidney beans, black beans, red lentils)	• Chicken • Eggs • Shellfish • Wild salmon • Tuna • Legumes/lentils (such as butter beans, green and puy lentils)	• Tofu, tempeh • Cod • Sea bass • Prawns • Eggs • Legumes/lentils (such as green and puy, haricot beans)	• Tofu, tempeh • Eggs • Wild salmon • Chicken • Halibut • Grass-fed beef • Legumes/lentils (such as borlotti and chickpea, red lentils, black beans)

	MENSTRUAL PHASE	FOLLICULAR PHASE	OVULATORY PHASE	LUTEAL PHASE
GRAINS	• Buckwheat • Brown rice • Wild rice • Oats	• Oats • Barley • Rye • Wheat • Quinoa	• Quinoa • Amaranth	• Brown rice • Black rice • Millet • Buckwheat
NUTS AND SEEDS	• Walnuts • Hazelnuts • Chestnuts • Peanuts • Ground flaxseed • Pumpkin seeds	• Brazil • Cashews • Pine nut • Ground flaxseed • Pumpkin seeds	• Almonds • Pistachio • Sunflower • Sesame seeds	• Walnuts • Peanut • Pistachio • Sunflower • Sesame seeds
HERBS AND ADAPTOGENS *tea infusions/ tinctures/ fresh herbs to add to meals	• Nettle • Ashwagandha • Turmeric • Fennel • Chamomile • Chaga mushroom	• Nettle • Dandelion • Milk thistle • Holy basil • Schisandra • Parsley • Basil • Lion's mane mushroom	• Mint • Coriander • Dill • Parsley • Red clover • Shatavari • Maca • Dandelion • Cordyceps mushroom	• Dandelion • Rosemary • Ashwagandha • Cinnamon • Ginger • Burdock root • Chamomile • Reishi mushroom
EXTRAS	• Miso • Tamari • Sea salt • Cacao • Collagen • Grass-fed butter • Extra-virgin olive oil • Bone broth • Dark chocolate (over 70%) • Red raspberry leaf tea • Decaf green tea • Ginger and turmeric tea with local or manuka honey (see my immunity tea recipe on page 170)	• Nut butter • Apple cider vinegar • Flax oil • Avocado oil • Matcha • Green tea	• Extra-virgin olive oil • Pickled onions (see page 200) • Pickled vegetables • Kombucha • Mint tea	• Spirulina • Dark chocolate (over 70%) • Sea salt • Grass-fed butter • Bone broth • Avocado oil • Matcha • Green tea (part 1) • Ginger and turmeric tea (part 2) with local or manuka honey (see my immunity tea recipe on page 170) • Cacao

	MENSTRUAL PHASE	FOLLICULAR PHASE	OVULATORY PHASE	LUTEAL PHASE
EXCERCISES	Your hormones are at their lowest point, so make rest and fresh air a priority. Swap your exercise regime for a sound bath, breathwork and/or meditation. In terms of movement, choose walking in nature, stretching and gentle yoga.	Oestrogen is increasing, giving you more energy and motivation to work out. Focus on strength training and increasing high-intensity workouts as your energy increases each day and as you move towards the ovulatory phase.	LH is rapidly rising and as oestrogen peaks it couldn't be a more ideal time to push yourself and embrace your body's resilience with high-intensity cardio, such as cycling, interval training and kickboxing, while also focusing on building muscle mass, as the surge of testosterone helps with the growth, maintenance and repair of muscles – go for it!	Part 1: continue with cardio if you feel like it, but include strength-based exercise, too. Part 2: as progesterone is at its highest it's time to take it down a notch, prioritising slower, flexibility-based movements, such as Pilates, barre and swimming – but most importantly, resting when your body is telling you it needs it.

When trying cycle syncing, keep an open mind and understand that individual experiences vary. Always listen to your body, be open-minded in your approach and consult with a healthcare professional for tailored guidance and one-to-one support.

A WORD ON CYCLE SYNCING: IT DOESN'T HAVE TO BE PERFECT!

Please note the above guidance is aimed at giving your body a nourishing, structured way of eating that supports your fluctuating hormones and feminine energy as much as possible. But it is just that – guidance. So if you want to eat something that's recommended in a different phase, that's ok, too, even just being mindful about the energetics of the food you eat (e.g. cooked or raw) and eating a few of the key suggested foods within each phase can make a positive impact. It's not about perfection.

HOW TO AVOID BLOATING

Feeling bloated during your luteal phase? Try sleeping on your left side, it can help with bloating and digestion because gravity helps move food and waste through your system. The stomach's natural position is on the left side, where it can digest food more effectively. This simple hack can also help prevent heartburn and keep pressure off your internal organs.

Seed Cycling for Hormone Balance: All You Need to Know About Using Food as Medicine

Seed cycling is for you whether your hormones are out of balance, your cycle is irregular, heavy or light, you want to optimise your fertility, improve PMS or you are peri- or post-menopausal.

WHAT IS SEED CYCLING?

Seed cycling is a naturopathic practice that involves eating and rotating specific seeds during the follicular and luteal phases of your menstrual cycle to promote the optimal balance of two of your primary sex hormones – oestrogen and progesterone.

Flax, pumpkin, sesame and sunflower seeds have been chosen for their hormone- and cycle-optimising properties via the balancing actions of phytoestrogens, zinc, selenium and vitamin E. As a bonus, they are also great sources of fibre, manganese, magnesium, copper, thiamine, omega 3 (and in lesser amounts omega 6), essential fatty acids and healthy fats, which are the building blocks for your hormones and promote blood flow to the uterus, increase progesterone secretion, and help maintain healthy cell membrane for enhanced reproductive health. Plus, the seed hulls contain phytoestrogens called lignans, which bind to excess hormones, to regulate and achieve a balanced hormonal system.

Seed cycling is a perfect example of using food as medicine, not just to support hormonal health but to take it to the next level, and the way I see it, there is everything to gain and nothing to lose by boosting your diet with these high-fibre, nutrient-dense seeds. There really are no negatives.

Seed cycling has been around for hundreds of years and is a practice of which I'm a huge advocate. I feel passionate about it because of the real-life results I've seen, in my own hormone-balancing journey (it was one of the nutrition interventions that contributed to regulating my cycle and, eventually, healing my body of PCOS symptoms), and also the clients I've had the privilege of treating.

HOW DOES IT WORK?

The aim of seed cycling is to help regulate oestrogen in the first half of your cycle and progesterone in the second half.

Starting with the follicular phase (days 1–15 of a twenty-eight-day cycle or from new moon to full moon, if your cycle is irregular or non-existent and you are following the moon cycles), including your menstural phase and ovulatory phase, you consume 1 tablespoon each of ground flaxseed and pumpkin seeds every day until you ovulate, to gently and naturally regulate oestrogen levels (improving levels and preventing an excess). Pumpkin seeds are high in zinc, which supports progesterone production and release as you move towards the second phase of your cycle, so if you are not ovulating (or not ovulating every month), this will help.

During your luteal phase, which starts after ovulation (days 16–28 or from full moon to dark moon), you consume 1 tablespoon of sesame and sunflower seeds every day to boost progesterone and prevent an excess of oestrogen. See how these seeds work in synergy to create balance? It can often be the missing piece of the hormone balance puzzle.

HOW CAN SEED CYCLING HELP ME?

When your hormone levels are balanced, oestrogen rises during the first half of your cycle. During the second half, progesterone rises while oestrogen slowly declines.

An imbalance between oestrogen and progesterone can contribute to PMS symptoms, menstrual cramps, acne, short luteal phases, anovulation, irregular cycles and amenorrhea. Seed cycling can naturally and effectively treat common symptoms of imbalanced oestrogen and progesterone, including PMS, irregular periods, PCOS and endometriosis to name just a few. It's also known to ease symptoms of menopause and boost fertility.

But even if you are not experiencing any of the above, you can still benefit from seed cycling and lots of people do, including me. Even though my cycle is now regular and healthy, I still turn to seed cycling when I know my hormones could do with a little extra TLC (say, if I notice low mood or am feeling generally not myself). Because not every month is the same, it's important to maintain hormone balance by using foods to help your body when needed and seed cycling is a proactive and targeted way to do this.

HOW DO I INCORPORATE THE SEEDS INTO MY DIET?

These seeds might be small, but they are mighty and easily incorporated into your diet with minimal faff. Simply sprinkle them over your breakfast, lunch or dinner, add them to smoothies or energy balls or crackers (see page 48 and 55) or stir into yoghurt or kefir. To make it quick and easy for me to use the seeds, I combine them as needed for the follicular and luteal phases of my cycle, pour them into two airtight glass containers and keep them in the fridge to retain their natural oils and nutrients. I even had cute labels printed for the jars to make it quick and easy to identify and grab them in the morning when I want to sprinkle them over my brekkie.

I recommend buying organic seeds and using a coffee grinder or food processor to grind them down into powder. Avoid buying pre-ground seeds, as they oxidise quickly and lose their hormone-balancing superpowers.

FOLLICULAR-PHASE SEED CYCLING: WHY FLAX AND PUMPKIN?

FLAXSEED contains lignans, which mimic the action of oestrogen and modulate levels to promote balance.[23] The fibre in flaxseed can also help your body eliminate excess used hormones that are secreted into the intestine for removal via your stool. This is another way that seeds (especially flax for their high-fibre content) are beneficial for hormone balance.

PUMPKIN SEEDS are rich in zinc, a key mineral for supporting reproductive health, particularly when it comes to promoting production of progesterone, which is released in the second half of your cycle (luteal phase). Incorporating pumpkin seeds during the follicular phase of your cycle can support optimal levels of progesterone, which has been linked to healthy ovulation (ovulation is so important, as we've already learned – see page 30) and, therefore, increased fertility.[24,25]

Consume 1 tablespoon ground flaxseed and 1 tablespoon pumpkin seeds to support oestrogen balance.

LUTEAL-PHASE SEED CYCLING: WHY SESAME AND SUNFLOWER?

Once you've ovulated (read below for guidance on what to do if you don't), switch your daily seeds to sesame and sunflower for fourteen days during your luteal phase, until day 1 of your period (regardless of whether or not your period starts). The seeds you

LUTEAL PHASE
SESAME & SUNFLOWER SEEDS
FOLLICULAR PHASE
GROUND FLAX & PUMPKIN

consume during the luteal phase in the lead-up to your period aim to support healthy progesterone levels.

SESAME SEEDS are another rich source of zinc, which supports progesterone production to promote a healthy luteal phase and contains compounds, which bind to excess oestrogen to further promote balance of the two hormones. They're also a great source of vitamin E, which acts as a powerful antioxidant, dampening down inflammation, along with your PMS symptoms, too.[26] Sesame seeds have also been seen to increase sex-hormone-binding globulin (SHBG), which is lower in women with PCOS.[27]

SUNFLOWER SEEDS contain selenium and vitamin E, which supports the functioning of the corpus luteum – a temporary endocrine gland that forms from the emptied ovarian follicle after ovulation and are rich in vitamin E, which also promotes healthy progesterone production during the luteal phase.[28] Selenium also helps detox the liver of excess oestrogen[29] and is also essential for optimal thyroid function, which, in turn, is vital for healthy hormones and regular cycles.

Consume 1 tablespoon sesame seeds and 1 tablespoon sunflower seeds to support progesterone balance.

HOW TO GET STARTED

Option 1: the 'standard' seed cycle for cycles between twenty-four and thirty-four days

If your cycle falls somewhere between 24 and 34 days, I recommend following the seed-cycle guidelines explained above, which are based on a 'standard' 28-day cycle, regardless of whether your cycle is 28 days or not, or if you ovulate perfectly on day 14. The idea is that the rotation of the specific seeds will gently help to nudge oestrogen and progesterone into place and will regulate your cycle. The consistency of the 14-day seed rotation helps bring hormones into balance, encouraging optimal menstrual-cycle rhythm.

So to get started, follow the follicular-phase guidelines from day 1 of your cycle (the first day you bleed). Continue for 14 days, then on days 15–28, switch to the luteal-phase seeds, whether you have ovulated or not. If your cycle falls outside of these parameters, follow option 2, below.

Option 2: the moon seed cycle for irregular cycles and amenorrhea

If your cycle is long and irregular or you are not currently menstruating, follow the cycles of the moon instead of your menstrual cycle. It's a lot easier than it sounds, I promise.

The moon is powerful; it controls the waves in our oceans, impacts gravity and has a significant impact on your cycle. It's no coincidence that the moon phases (from new moon to full moon to new moon) occur on a 28-day cycle. Cycling seeds with the phases of the moon can help get your body back into the rhythm of a 28(ish)-day cycle.

TIP: I recommend googling the moon-cycle phases or using an app such as Moonly.

If you are following the moon seed-cycle guidelines, you will start consuming your daily seeds on a new moon and would treat this new-moon phase as your follicular phase. After day 14 of the moon cycle, when a full new moon occurs, you switch over to the luteal-phase seeds, which would last until day 28, when another new moon occurs.

FAQs

Q. How long until I notice results?
A. Many women experience the positive impacts in the first month or two of daily seed cycling aligned to cycle syncing or eating a healthy, diverse, balanced diet. However, it can take four cycles for your hormones to be regulated, as that's how long it takes for a follicle to mature and be released at ovulation, so consistency is key. (Please note, results are individual and you cannot seed cycle your way out of an unhealthy diet.)

Q. What if I'm oestrogen dominant?
A. The clever thing about phytoestrogens is that they naturally increase or decrease oestrogen, depending on what the body needs, which research shows can promote regular menstrual-cycle length.[30,31] This seems to be down to the phytoestrogens found in flaxseed and the way they not only modulate the activity of enzymes between the balance of oestradiol and estrone but the fact they are high in fibre, too, which means they are a good way to bind to excess oestrogen and excrete it via your poo! All in all, according to the science, there is nothing to fear when it comes to consuming flaxseed if your oestrogen levels are on the high side as it can be an effective way to regulate oestrogen levels (along with the niggling associated symptoms).

Q. What if I'm going through perimenopause, menopause or post-menopause?
A. You will still feel the benefits of seed cycling and cycle syncing, even if your cycle is in the process of slowing down/stopping (perimenopause) or has completely stopped (menopause and post-menopause). Since levels of oestrogen and progesterone naturally decline during menopause, seed cycling can provide extra support to alleviate symptoms associated with lower hormone levels. If this applies to you, instead of tracking your cycle using an app, track the moon cycles and eat aligned to nature's seasons. Many women have reported a significant decrease in their menopause symptoms, such as hot flushes, mood irregularities, migraines and more.

Q. What if my periods are irregular or I have PCOS?
A. Scientific research shows positive results of seed cycling to manage and reduce PCOS symptoms.[32] In a study, approximately 36 per cent of participants experienced significant changes or complete degeneration of cysts. Additionally, nearly half of the participants had fewer cysts compared to a control group where cyst numbers increased.[33]

Q. Will seed cycling and cycle syncing support my fertility?
A. What you eat has been proven to have a hugely positive impact on regular cycles, fertility and egg quality.[34,35] Eating aligned to your cycle to support hormone balance and seed cycling, too, will promote healthy hormones and encourage ovulation.

Q. Can I start seed cycling at any time?
A. I recommend starting on day 1 of your cycle to create rhythm (plus, it's easier to keep track).

Q. What if I miss a day?
A. Consistency is key, but don't worry if you forget; just continue the next day.

Q. What if I'm on hormonal contraceptive?
A. I still recommend seed cycling while on the Pill. The nutrients in the seeds are very beneficial and the fibre in them is great for your gut health, too, so there is no reason not to do it. However, you will experience the best hormone-balancing results when coming off the Pill to regulate your natural cycle and to prevent negative post-Pill symptoms, when your body is free of synthetic hormones.

Q. Can I do seed cycling while pregnant?
A. There is no specific benefit to following the seed-cycling guidelines if you are pregnant, but I'd suggest incorporating a combination of all of the seed-cycling seeds into your diet to increase nutrients and fibre during your pregnancy.

Q. Can I seed cycle while breastfeeding?
A. Seed cycling can be an effective way to support your body post birth and during breastfeeding. It can help by providing your body with the essential nutrients it needs to for hormonal balance and overall wellbeing.

Q. Is seed cycling safe?
A. Seed cycling provides benefits due to the increase in hormone and gut-supportive nutrients, and there is no medical reason or research to suggest that it causes adverse effects on health. However, if you suffer with IBS, you should consult your GP, due to the increase in fibre.

Q. What evidence is there that seed cycling works?
A. Scientific research on the impact of seed cycling is limited because, sadly, women's health (especially when it comes to holistic medicine and practices) is not a priority. But this does not mean it doesn't work, and scientific research backs up the nutritional profiles of the targeted seeds and their positive impact on hormonal health during the different phases of the menstrual cycle.[36] The cynic in you might want to believe that these tiny seeds couldn't possibly be the solution to your irregular cycles and PMS, but you need only google them or search Instagram or TikTok for thousands upon thousands of positive reviews. You have everything to gain and absolutely nothing to lose by incorporating seed cycling into your daily lifestyle.

Remember, seed cycling is just one piece of the Nourish Method Cycle Syncing hormone-balancing puzzle. Eating healthy wholefoods, getting enough nourishment at each meal and limiting the endocrine disruptors (see page 113) all play an important role, too. Nothing about hormone balancing can be actioned in isolation: nourish the whole of you to balance the whole of you.

In the next chapter we'll take a deep dive straight into exactly how to start supporting your body to optimise your hormones, starting with your second brain: your gut.

Cycle-craving Recipes

'The cure is in your kitchen.'

Seed-cycling Energy Balls

These hormone-balancing, no-bake seed-cycling energy balls are a great way to get your daily hormone-balancing seeds into a tasty nutrient-dense snack and will help you to create a seed-cycling habit that sticks, keeping you satisfied.

The seeds aligned to the phase of your cycle will promote the optimal balance of oestrogen and progesterone, but this recipe also contains ground almonds and collagen to support your skin, and natural sweetness from the dates to satisfy sugar cravings, with a decent dose of B vitamins and fibre to keep you energised and nourish your microbiome.

These energy balls keep well in the fridge for seven days and freeze well, too.

Makes 7 large energy balls (1 ball per day) or 14 mini energy balls

For the recipe base
100g nut butter (almond or peanut)
60g ground almonds
20g collagen or protein powder
150g Medjool dates, stones removed
Pinch of sea salt
1 tbsp coconut or MCT (medium-chain triglyceride) oil
¼ tsp ground cinnamon

For the follicular phase
7 tbsp flaxseeds
7 tbsp pumpkin seeds

For the luteal phase
7 tbsp sesame seeds
7 tbsp sunflower seeds

1. Pour the relevant seeds for the phase of your cycle into a food processor and blitz.
2. Add the rest of the ingredients and blitz to combine.
3. Roll the mixture into balls, then chill in the fridge for an hour before sitting down to enjoy with a cup of tea.

TIPS: If you prefer eating whole seeds, you can roll the recipe base mix in the seeds for your phase instead of blitzing them.

Add 35g cacao powder to the follicular-phase batch to replenish lost stores of iron from your period, plus magnesium to prevent PMS cramping.

Cacao and Collagen Bites

These hit the sweet spot, without refined sugar and with all the added hormone-nourishing extras, including magnesium, iron, amino acids/protein (from the cacao and collagen), proven blood-sugar-balancing benefits (from the slightly spicy ground cinnamon), plus gut-healthy fibre and antioxidant-rich vitamin E (from the ground almonds) to support your skin health. They will keep in the fridge for up to seven days and can be frozen.

Makes 9 bites

3 tbsp cacao
3 tbsp collagen
100g organic gluten-free oats
150g Medjool dates, stones removed
40g ground almonds
Pinch of sea salt
2 tbsp coconut or MCT (medium-chain triglyceride) oil
½ tsp ground cinnamon
100g peanut butter

1. Add all the ingredients to a food processor and blitz until combined.
2. Roll the mixture into 9 equal-sized balls. Leave to set in the fridge for 1 hour before eating

Frozen Fruit and Nut Greek Yoghurt Heart Bites

When you fancy ice cream during your ovulatory phase or during the summer months of the year, but don't want the refined sugar, try these! They are made with juicy antioxidant-loaded berries and crunchy pistachios (one of the richest sources of vitamin B6, important for blood-sugar regulation and the formation of haemoglobin) to support your adrenals.[37] And they are drizzled with delicious dark chocolate and sprinkled with a tiny bit of sea salt, which, in my opinion, absolutely makes them.

Makes 6–8 hearts

180g Greek yoghurt
100g raspberries
100g pomegranate seeds
2 heaped tbsp chia seeds
100g pistachios, finely chopped (or mixed nuts also works)
3 tbsp raw honey
80g dark chocolate (70% cocoa solids)
Pinch of sea salt

1. Put the Greek yoghurt, berries, pomegranate, chia seeds and finely chopped pistachios in a bowl, along with the honey and give it all a good mix.
2. Line a tray with baking paper and spoon the yoghurt mixture on to it, creating 6–8 heart shapes with the back of the spoon.
3. Pop the tray in the freezer for 1 hour, until the hearts are set.
4. Melt the chocolate in a heatproof bowl set over a pan of boiling water over a medium heat.
5. Remove the yoghurt hearts from the freezer and either drizzle the chocolate over them or dip the tips into it.
6. Sprinkle with a little sea salt and then place them back on the lined baking tray and return it to the freezer to set for around 30 minutes. Store in the freezer for 7 days.

Berry and Ginger Kefir Compote

Berries really are nature's sweets. I've been drumming this into Sebby's little head since he could say the word 'snack', and once he believes this I'll be one happy mama! But seriously, berries are so sweet, delicious and bursting with goodness to help reduce inflammation and feed your skin from within. They are especially nutrient-dense when they're in season, so when they are you should embrace nature's offerings and make this gorgeous berry and ginger compote for when you need a boost. The probiotic-rich kefir will support your gut microbes and are great for kids, too!

Another idea is to serve the compote on a slice of sourdough, along with some nut butter for nature's version of peanut butter and jelly, or serve topped with pistachios, pecans, chia and hemp seeds (or your seed-cycling seeds) for a health-enhancing snack or dessert.

Serves 4

400g mixed berries
2½cm piece of fresh ginger, grated
Juice of ½ lemon
1 tsp grated lemon zest
4 tbsp maple syrup or raw honey
400g kefir yoghurt

1. Add the berries to a pan over a medium-low heat
2. Add the ginger, lemon juice and zest and maple syrup (or honey).
3. Simmer over a low/medium heat for approximately 10–15 minutes, stirring often (don't allow it to boil or caramelise too much).
4. Leave to cool before stirring in the kefir yoghurt. Transfer into a glass, airtight container and keep in the fridge for up to 7 days or transfer into ice-cube trays and freeze for later.

Rosemary and Sea Salt Seed-cycling Crackers

If you make one recipe in my book, please make it this one... I first made these versatile seeded crackers with my lovely mother-in-law (Mummy Shand) and we've been crunching our way through them as a family ever since. (Sebby is also obsessed and loves them with nut butter on top!)

I now make a weekly batch and incorporate them into breakfasts, lunches and snacks because they really are the ultimate hormone-boosting base to layer your toppings onto, whether it's eggs, hummus, tuna or chicken, or into my protein-rich Fresh Herby Tzatziki (see page 58).

Serves 1

For the follicular phase

160g flaxseed (not ground)
140g pumpkin seeds
5 tbsp chia seeds
1 tbsp rosemary, fresh or dried (or swap to your fave herb)
½ tsp sea salt
200ml boiling water

For the luteal phase

180g sunflower seeds
120g sesame seeds
5 tbsp chia seeds
1 tbsp rosemary, fresh or dried (or swap to your fave herb)
½ tsp sea salt
200ml boiling water

1. Add the seeds and seasoning to a large mixing bowl. Add the water and stir well. Set aside for 15 minutes to allow the mixture to thicken to a gooey consistency.
2. Preheat the oven to 150°C.
3. Line a roasting tray with parchment paper and transfer the mix to the tray, smooth out evenly and press down using a fish slice – this makes them nice and thin.
4. Pop the tray on the bottom shelf of the oven and roast for 30 minutes on each side. When you flip them over use a knife or pizza cutter to cut into even sizes (if not, just break them-up later on). Allow to cool before tucking in!

Don't forget to adapt the seeds you use in this recipe to align with where you're at in your cycle. If you're not seed cycling, use either of the four seeds listed or a mix.

TIP: Mix ½ tin tuna, 1 tsp sweetcorn and ½ finely chopped spring onion with 1 tbsp live yoghurt, a squeeze of lemon and a sprinkle of chopped basil. Season with a pinch of oregano, sea salt and pepper, then spread on top of your seeded cracker for a deliciously nourishing lunch!

Cranberry and Pecan Cookies

These soft, crumbly, super-satisfying cookies are a firm favourite in our household. Even my husband, James, rates them, which is saying something! High in wholegrain fibre, they give all the cookie vibes without the unnecessary chemicals, making them an ideal way to curb your intense cycle sugar cravings.

Makes 6 cookies

200g oats
400g smooth peanut butter
½ tsp vanilla extract
1 tsp ground cinnamon
200ml maple syrup or raw honey
1 egg, beaten
60g dried cranberries, chopped
100g pecans, chopped

1. Preheat the oven to 180°C. Line a baking tray with parchment paper.
2. Place the oats, peanut butter, vanilla extract, ground cinnamon and honey in a big mixing bowl, along with the egg and mix well.
3. Add the chopped cranberries and pecans and combine well.
4. Drop large spoonfuls of the mixture on to the lined baking tray, flattening them down so they crisp up perfectly in the oven.
5. Bake for around 15 minutes or until they're lovely and golden.
6. Allow the cookies to cool and harden before tucking in. Keep in a glass, airtight container for up to 4 days.

Protein-rich Fresh Herby Tzatziki

I love this fresh and zesty dip, it's such a yummy way to add protein probiotics and micronutrients to your snack or spread into flatbreads or wraps. Adapt it to use the herbs you have in stock; I love it most with mint and parsley, but coriander works well too.

Makes 1 bowl

300g full-fat Greek yoghurt
½ cucumber, deseeded
2 garlic cloves, very finely chopped/minced
2 tbsp extra-virgin olive oil
1 tsp grated lemon zest
2 tbsp mint, finely chopped
2 tbsp flat-leaf parsley, finely chopped (or try coriander if serving with curry)
2 pinches of freshly ground black pepper and 1 generous pinch of sea salt (or to taste)

1. Combine all the ingredients in a bowl, tasting and adjusting the seasoning to your preference.

Hybrid Butter–bean 'Hummus—Guacamole'

The hybrid guacamole you didn't know you needed! Hummus and guacamole – can you think of a better combo? A pot of this will provide plant protein, healthy fats and fibre to renourish your system as a hormone-healthy snack. Dip in crunchy raw vegetable sticks, or spread over one of my seeded crackers (see page 55) when you need to bridge the gap between meals. Or layer into a sourdough sandwich and thank me later!

Makes 1 bowl

400g tin butter beans, drained and rinsed
½ avocado
1 garlic clove
5 tsp extra-virgin olive oil
Juice of 1 lime
Big pinch each of sea salt and freshly ground black pepper

1. Add all the ingredients to a food processor and blitz, until you have the desired consistency. And that's it!

Classic Guacamole

Makes 1 bowl

2 ripe avocados, mashed
2 spring onions, finely chopped
Small handful of cherry tomatoes, finely chopped
Juice of 1 lime
Handful of coriander, finely chopped
1 tbsp extra-virgin olive oil
Pinch each of sea salt, freshly ground black pepper and dried chilli flakes

1. Add all the ingredients to a food processor and blitz, until you have the desired consistency.

PMS-Free Pick-'n'-mix Bowl

During the latter part of my luteal phase, a few days before my period or at the start of it, I crave some sweet and saltiness and this little combo hits the spot without fail!

The showstoppers of this pick-'n'-mix bowl are the squidgy, melt-in-your-mouth Medjool dates – because not only are they deliciously sweet (I call them nature's caramel!), but they have lots of hormone health benefits. They're high in antioxidants (flavonoids) and studies show they help to lower inflammatory markers, making them the perfect healthy sugar substitute without the damaging effects of fake, refined sugar.[38] Dates are also high in fibre and can help to prevent menstrual-phase constipation and keep your bowels moving.[39]

A good dose of nutrients, such as potassium, magnesium, copper, B vitamins and iron, all of which are helpful to restore depleted nutrient stores during the luteal and menstrual phases of your cycle, when your reproductive system is working extra hard for you and therefore pulling on your nutrient stores.

Feel free to play around and switch the fruit to whatever you fancy. And if you are constipated, add a kiwi with the skin on to get your bowels moving.[40] It works like magic!

Serves 1

1–2 Medjool dates
1–2 tsp peanut butter
Pinch of sea salt
A few berries
A few walnuts, cashews or mixed nuts
1 square dark chocolate, over 70% cocoa solids (optional), chopped

1. Score the Medjool dates with a sharp knife, remove the stones and fill with a little peanut butter and a pinch of sea salt.
2. Add to a bowl, along with the berries, nuts and dark chocolate, if you fancy it.

TIP: For an extra chocolate hit, dip the peanut butter-stuffed dates into melted dark chocolate and freeze to set. Allow to come to room temperature before eating. It makes them taste like caramel truffles!

Peach and Mango Kefir Ice Lollies

Naturally sweet and so delicious, full of antioxidants and probiotics (the bacteria remains in tact but don't leave them in the freezer longer than a few weeks for maximum probiotic benefits) these gut-loving ice lollies make the perfect chemical and refined sugar free ice lollies for you and the little people in your life. Or if you're not up for a lolly just serve as a smoothie – it works just as well!

You'll need ice lolly moulds and a blender for these.

Makes 6–8 lollies

250ml kefir
2 ripe peaches, roughly chopped
150g mango chunks
2 tbsp chia seeds
250ml mango juice
2 tbsp raw honey or maple syrup (to taste)

1. Pour the kefir along with all of the other ingredients into a blender, blitz until smooth and transfer into the ice lolly moulds.
2. Pour any leftovers into a glass and drink as a smoothie. Or freeze the lollies overnight and use within 1–2 weeks.

Secret Squash Chocolate Mousse

Natalie (aka Twinny), one of my most cherished friends, made me a little pot of her squash chocolate mousse when I escaped to her house for some peace while writing part of this book, so I had to include it here, as it will forever remind me of that special time!

I remember it like it was yesterday. It was the first day of my period and I was craving something sweet but nourishing – and this did the trick! It's light, chocolatey and fudgy in texture and it's hard to detect the squash. It's become a firm favourite for me and for Sebby (and I love that it's another sneaky way to get more veggies into his tummy). And the great thing is, the fibre from the squash acts as a buffer to the sugar from the chocolate, making it a great cycle-craving dessert.

For the most delicious results, use the best-quality chocolate you can get your hands on.

Serves 4–6

300g butternut squash, peeled, deseeded, and cubed
150g dark chocolate (70% cocoa solids)
Pinch of sea salt (optional, but I prefer it with)
Drizzle of peanut butter
Handful of pistachios, crushed (or any nuts of your choice)
1 tbsp hemp seeds
1 tbsp chia seeds

1. Place the prepared butternut squash in a medium-sized pot and fill with water. Cook over a medium heat for approximately 20–30 minutes until completely soft.
2. Break the chocolate up and add to a bowl placed over a pan of boiling water and heat until melted.
3. Drain the squash and then transfer to a blender with the melted chocolate and sea salt (if using) and blitz, until silky smooth. (If you're making this for kids, add the sea salt when you serve your portion.)
4. Serve in a bowl with a drizzle of peanut butter, crushed nuts and sprinkle with the hemp and chia seeds. Store in a glass, airtight container and keep in the fridge for 4 days (it can also be frozen).

4. The Gut and Digestion

'Good gut health is good hormonal health.'

Your gut produces hormones that regulate your cycle, mood, hunger, sleep and stress – all the stuff you came here to improve, right? So I want you to use this chapter as your go-to and your starting point whenever your hormones are telling you they need your support.

The Gut Basics

Good gut health comes from a diet that is rich in a diverse range of high-fibre, colourful plants containing plenty of phytonutrients and pre- and probiotics. These gut-feeding heroes quite literally nourish your gut microbiome and feed your microbes, helping them to multiply, creating a flourishing internal garden, which means a healthier hormonal system. Let's learn how and what impacts our hormones from the perspective of our gut health and why daily bowel movements are key to keeping your hormones in balance.

GUT IMBALANCES AND HORMONE HEALTH

There are many factors that can cause an imbalance in your gut bacteria, ranging from poor diet (including a high intake of UPFs and refined sugars) to stress, environmental toxins and medication (such as antibiotics).[41] These can irritate the lining of the gut, causing inflammation, nutrient deficiencies and a range of health problems, including hormone imbalances.

If your gut is imbalanced and you are suffering with digestive difficulties, this might indicate you are not extracting and absorbing the nutrients from your food, causing potential nutrient deficiencies mentioned above, and it might also mean that you are not excreting hormones and toxins from your body, which would also contribute to hormone imbalances. If you want to understand how to nourish your gut to help your hormones, you are in the right place.

Your gut is the control panel to how your health needs to be prioritised daily through nutrition. If your gut is thriving, you will be helping your hormones to thrive, too. Nourishing your gut microbiome via the food you expose it to is also the gateway to supporting all your body systems, helping you with any health goal you have, from weight management, energy levels, glowing skin and immunity to, of course, hormone metabolism and overall balance.

WHAT IS THE GUT?

Your gut is your gastrointestinal system (GI) and includes your stomach, intestines and colon. It's a series of organs joined in a tube that runs from your mouth to your anus, designed to absorb, break down and digest food into its component parts (amino acids, sugars, fats, fibre, vitamins, minerals and phytonutrients), which are then utilised in your body to regulate and maintain health. As well as absorbing and digesting, part of your gut's job is to keep out harmful microbes and toxic metabolites (small toxic molecules) that cause inflammation and contribute to many diseases.

WHAT IS THE GUT MICROBIOME?

Your gut microbiome refers to all the microorganisms (aka bacteria) that live within your gut garden. As you know, to grow beautiful flowers and tasty herbs, you must water and nourish the soil. And it's the same when it comes to bettering your gut health: you need to feed it well if you want the good bugs that are integral to good health (and therefore good hormonal health) to grow. Your gut's favourite foods are colourful wholefoods that are brimming with vitamins, minerals and antioxidants (i.e. phytonutrient-rich plants) and high in pre- and probiotics to promote the good bacteria that help it to thrive, so keeping you healthy and balanced.

Every single time you eat, you are either feeding the good bugs or the bad bugs; you either support and promote the healing of your gut lining or you expose it to foods that can cause leaky gut (an unhealthy gut lining, which allows bacteria and toxins to 'leak' through the intestinal wall and is linked to diseases such as gut dysbiosis, immune-system imbalances, IBS and nutritional deficiencies).[42]

'GUT INSTINCT'

The age-old term we all use so readily – 'gut instinct' – was coined for a reason. Listening, and I mean really listening, to your gut instinct is the cornerstone of intuitive eating, as your gut is your second brain (no, really – it is) and, as such, it heavily influences pretty much every system in your body, including, of course, how balanced and happy your hormones are.

Whether you choose to believe it or not, your gut is *always* telling you what it needs, but you may not always remember to tap in and listen to what it's saying (because ... life!). Or you may simply not know which ingredients to feed it to help it thrive. But either way, I hope this chapter encourages you to nourish your gut to better health by integrating the plant-diverse recipes I've created with your gut microbes in mind. Because when you put your gut microbes first via the food you feed them with, you are serving your hormones and helping your body and brain to thrive. Basically, it always starts with the gut.

HOW WILL I KNOW IF MY GUT NEEDS EXTRA SUPPORT?

You might be experiencing a multitude of overlapping symptoms manifesting as low energy, low mood and sugar cravings in conjunction with *regular* bloating (the odd bit of bloating can be totally normal, especially after a high-fibre meal), constipation, loose stools, stomach 'growls' or generalised gut 'discomfort', which will likely worsen when you feel stressed.[43]

I want you to start observing these symptoms (which may have become seemingly 'normal' to you), as your gut's way of saying it needs something to change; it needs more than you're currently giving it. And this is likely to be in the form of its favourite type of food: fibre. When you eat fibre, something called short-chain fatty acids (SCFAs) are produced, which feed the good guys (good bacteria) living in your gut, so they have an opportunity to multiply and thrive. And as we already know, a thriving microbiome is going to help your hormones do their jobs properly.

TIP: Be curious

Question your normal. Just because bloating is something you are used to, and perhaps lots of other people suffer from it, too, that doesn't make it ok. Normal is not optimal, and normal is not the goal.

The Gut–Hormone Connection

Your gut produces more neurotransmitters (chemical signals used by the nervous system) than your brain – a startling 95 per cent of your serotonin (the 'happy hormone') and 50 per cent of your dopamine.[44] And this is relevant to your hormone-healing journey because serotonin, this superhero neurotransmitter, plays a fundamental part in regulating your mood and mental and emotional health, and it's strongly connected to anxiety, insomnia and depression.[45] Plus, low serotonin levels are linked to constipation.[46] So you've probably got the message by now that your gut is responsible for more than just how well you digest food and how often you go to the loo.

Tryptophan is an essential amino acid needed for growth and many metabolic functions. It comes from food (or supplements) and our bodies convert it to 5-hydroxytryptophan, then serotonin, then melatonin. Melatonin helps to regulate the sleep–wake cycle, and serotonin helps to regulate appetite, sleep, mood and pain here too.[47] The liver can also use tryptophan to produce niacin (vitamin B3), which is needed for energy metabolism and DNA production. If your gut microbiome is lacking in serotonin-supportive bacteria, you can imagine how this could directly impact how happy, motivated and mentally stable you feel. In fact, scientific research suggests that diets low in tryptophan decrease serotonin levels.[48] And low serotonin levels equal low mood.

One of my aims in this chapter is for you to naturally encourage your microbes to produce more serotonin from the food you feed your gut with to improve your mood, mental health and cognition (see my high-fibre, gut-nourishing, tryptophan-rich recipes on pages 72–109, for support with this process). Tryptophan-rich foods include oats, bananas, dried prunes, cacao, dark chocolate, milk, tuna, cheese, wholegrains, chicken, turkey, peanuts, pumpkin and sesame seeds, tofu and soy.[49] In addition, eating complex carbs in your evening meal supports your sleep hormones by blunting cortisol, raising serotonin and increasing GABA levels (an anti-anxiety neurotransmitter) – all the good stuff!

The Gut–Oestrogen Axis

As well as being home to serotonin, your gut microbiome also houses a collection of bacteria (the oestrobolome) that metabolise and modulate your body's circulating oestrogen. It's the bacteria in your gut and the oestrobolome that affect how balanced your overall oestrogen levels are, which, in turn, can impact weight, libido and mood. Too much circulating oestrogen can drive PMS symptoms and hormone conditions such as PCOS, endometriosis and fibroids.[50] Low circulating oestrogen levels can present as irregular periods, dry skin, vaginal dryness and irritability.

Dysregulated oestrogen levels can impact your metabolism, energy levels and libido and can even make you more susceptible to certain diseases, most commonly including breast cancer (in cases of oestrogen excess) and osteoporosis, particularly in post-menopausal women (when levels are low – because oestrogen acts as a natural protector of bone strength).[51, 52]

TIP: Instead of fasting during the day, I recommend introducing a twelve-hour fast from when you finish your dinner until breakfast the next day to stimulate your parasympathetic nervous system. This activates something called 'rest and digest', giving your digestive system time to do its healing and internal housekeeping.

What's the science saying?

Studies have been carried out to investigate the microbial diversity of the gut microbiome and its impact on sex-hormone levels and related conditions and diseases, such as ovarian cancer, PCOS, endometriosis and osteoporosis.[53]

Having read hours', days', maybe weeks' worth of research papers on the intimate relationship between the microbiome and hormone balance, my findings not only confirm that the diversity of your gut does indeed influence levels of circulating sex hormones (including oestrogen, testosterone, progesterone and corticosteroids), but oestrogen most significantly.

Your gut microbiome regulates oestrogen, producing something called beta-glucuronidase, an enzyme which essentially frees up oestrogen for your body to use, helping you to function optimally. But when this process is impaired and your microbes are not able to produce enough beta-glucuronidase (or they produce too much), it can cause oestrogen to revert to its unconjugated active form and it is then absorbed back into the bloodstream, resulting in too much circulating oestrogen (aka oestrogen dominance/excess oestrogen), causing a whole host of undesirable hormonal symptoms and increasing your risk of oestrogen-driven conditions (including breast cancer, ovarian cancer, endometriosis and PCOS, which rely on oestrogen to develop and grow).[54,,55]

Dysregulation of the oestrobolome affecting circulating oestrogen levels is thought to be caused by your gut microbiome being low in bacterial diversity, due to a lack of fibre varieties, as well as low intake of pre- and probiotic foods. Don't worry, though, this can be naturally treated by eating a consistently healthy, balanced diet brimming with phytochemicals (natural bioactive compounds found in plant food that work with nutrients and dietary fibre to protect against disease) and by regulating how often and how 'well' you poo. This is an area of ongoing research, but the findings are undeniably significant and should motivate you even more to prioritise your gut health if you are serious about looking after yourself as a female on all levels, from menstruation through to menopause.

The Power of Poo

Constipation or loose stools indicate gut dysbiosis – a fancy word meaning your gut is out of balance. To reiterate, an imbalanced gut means imbalanced hormones (you get the picture now, right?) and you need to poo daily, ideally (or daily-ish, for some people), in order to maintain hormone equilibrium. This is because the hormones that have been 'used' and have carried out their jobs for you need to be excreted via your poo, and if they're not (due to constipation/irregular bowel movements), they are instead recirculated into your bloodstream and can cause many hormonal imbalances. So never underestimate the power of your poo – it says a lot about your hormonal health.

HOW YOUR CYCLE AFFECTS YOUR POO

It's very normal for your bowel movements to change during your cycle in alignment with the peaks and plummets of your hormones. Most commonly, it's in the lead-up to your period (luteal phase) and during your bleed (menstrual phase) that you will notice changes in your poo. This is a result of a rise in progesterone, which causes constipation, and then an increase in prostaglandins (a type of lipid compound) during your period. As well as causing the smooth muscle of the uterus to contract and shed its lining (your period bleed) and abdominal pains (period cramps), they can also cause loose stools and diarrhoea for some people, by stimulating the smooth muscle part of the colon to contract faster.[56,57]

Oestrogen (the queen is back!), unsurprisingly, has a part to play in your period poo, too, as it can cause your gut to absorb more water and salt from your poo as it moves through it. So if you are someone who experiences constipation during the start of your period, this is likely down to higher levels of oestrogen.[58]

While there is no 'normal' period poo, your menstrual cycle is deeply connected to and influenced by what you eat (and how well fed your gut microbes are) during the different phases and by the changing levels of your key sex hormones. This can impact how your digestive system behaves, how you poo and, of course, how you *feel* (hello, serotonin). Following The Nourish Method, with the special gut recipes I've created for you in this chapter, is an effective, holistic way to give your gut what it needs to create balance during your cycle.

YOUR BODY'S MASTER ANTIOXIDANT: GLUTATHIONE

Glutathione is the body's master antioxidant. Produced in your cells, it's comprised largely of three amino acids: glutamine, glycine and cysteine. Glutathione reduces oxidative stress (an imbalance between free radicals and antioxidants in your body) that can contribute to different chronic conditions, including autoimmune disease, as it works to protect cell mitochondria by eliminating free radicals (highly reactive and unstable molecules that are made by the body naturally as a by-product of metabolism).[59] Glutathione is abundant in the mucosal cells of the gastrointestinal tract. These cells, which make up the tract's mucosal lining, are essential for gut-barrier function. This protective barrier ensures that harmful substances don't leak into your bloodstream, while allowing essential nutrients to pass through. When glutathione is low, or there's a deficiency, it can lead to inflammation and increased intestinal permeability, aka 'leaky gut'.[60] Glutathione contains sulphur molecules, which may be why foods high in sulphur help to boost its natural production in the body. These foods include cruciferous vegetables (particularly broccoli sprouts), as well as onions, garlic, eggs, nuts, legumes and lean animal protein. Milk thistle and flaxseed also boost levels in the body. Glutathione is also negatively affected by insomnia. So getting enough rest on a regular basis can help to increase levels.

TIP: If you are constipated, one of the top tips and constipation hacks is to eat two kiwi fruits every day with the skin on (this is important), until your bowels are regulated.[61] Kiwis are high in two types of fibre – both soluble and insoluble – which work together simultaneously. As soluble fibre moves through the digestive system, it is able to hold water, creating larger and softer stools that move more easily through the colon (when stools are hard and dry, they have trouble moving through – that's when you experience straining); insoluble fibre also helps the movement of stools through the digestive tract by adding extra bulk. An optimal proportion of each fibre type results in a lubricating effect that helps to promote bowel movement.[62]

Gut-hormone nourishing foods

Think of prebiotic-rich foods as quite literally food for your gut bugs. Prebiotic molecules are plant fibres, so eating lots of plants in your meals is the best way to increase your prebiotic intake:[63] mushrooms, leeks, garlic, onions, asparagus, Savoy cabbage, oats, cacao, flaxseed, cashews, almonds, chicory root, Jerusalem artichokes, barley, apples, burdock root.

Probiotic-rich foods are living gut bugs (live microorganisms that increase the diversity of bacteria in your gut), helping to reduce the growth of pathogenic microbes in our gut, aid digestion, boost the helpful chemicals that our gut bacteria produce and primes your immune system:[64] live yoghurt, kefir, kimchi, sauerkraut, miso, tempeh, kombucha (avoid added sugars), certain cheeses, such as Parmesan, aged Cheddar and Swiss cheese.

Phytonutrients (natural chemicals or ones produced by colourful plants) support your gut bacteria and are associated with good hormonal health. They are rich in antioxidants, anti-inflammatory compounds and have neuroprotective properties:[65] berries, root veg (such as squash, sweet potatoes, parsnips, ginger, carrots), tomatoes, kale, walnuts, pecans, green tea, apples, broccoli, legumes, pak choy, artichokes, spinach, herbs, spices, dark chocolate (the darker, the more nutrients, the better).

The Nourish Method Gut Reset Ritual

Before you get your gut-feeding cooking under way, I want you to stop, sit down on a mat, tune into your body and perform an active reset to help aid digestion, bloating and get things moving.

Start by taking four nourishing, deep breaths, directing it into your gut area, hold for a count of four, then take one big detoxifying breath out. Now, slowly laying yourself flat on the mat, twist your torso to one side with your legs rested in the opposite direction, hold for a minute, then do the same on the other side. This helps to release stagnant energy in your gut and supports your digestive system. Reconnecting with your gut strengthens the gut–brain connection by instantly calming your nervous system. And it sends nourishing signals to your hormones that you care and good things are coming. This ritual will also strengthen your intuitive-eating muscle and positively impact the gut–hormone connection.

After this reset, take the opportunity to supply your gut with the foods it craves using the gut-nourishing recipes that follow and my intuitive-eating tips.

Gut and Digestion Recipes

Multivitamin Gut-nourish Bowl

This beautiful, diverse, high-fibre and high-protein recipe is quite literally a bowl of vitamins and minerals that work simultaneously to nourish your gut and feed your hormones. It is particularly rich in vitamin C from the pepper, tomatoes, pomegranate and leafy greens, quercetin from the red onion (known for its powerful anti-inflammatory effects) and plant-based omega 3 (ALA) from the walnuts, to top-up nutrient stores, especially if you are feeling fatigued and depleted.

It's a soul-feeding dish that not only looks good, but does so much good, too!

Serves 2

1 small–medium-sized butternut squash
1 tbsp extra-virgin olive oil
1 tbsp za'atar
Pinch each of sea salt, black pepper and dried chilli flakes (optional)
2 handfuls of spinach
2 small handfuls of rocket
1 Romano pepper
1 avocado
Big handful of cherry tomatoes, halved
¼ red onion, thinly sliced
120g tin tuna (in olive oil)
Handful of crushed walnuts
2 tbsp pomegranate seeds
Handful of flat-leaf parsley, chopped

For the nourish-bowl dressing

3 tbsp extra-virgin olive oil
1 tsp Dijon mustard
1 tsp white wine vinegar
2 tsp maple syrup or raw honey
Pinch each of sea salt and freshly ground black pepper

1. Preheat the oven to 200°C.
2. Peel and chop the squash into small chunks, drizzle with the extra-virgin olive oil and season with the za'atar, sea salt, pepper and chilli flakes. Roast in the oven for 30 minutes, turning halfway through cooking time.
3. Add the spinach, rocket, pepper, avocado, cherry tomatoes, red onion and tuna to a bowl.
4. Combine all the dressing ingredients.
5. When the squash is golden, add it to your salad bowl, then top with the crushed walnuts, pomegranate seeds and parsley and drizzle with the dressing.

'Feed-the-family' Rosemary Lentil Ragu

This recipe provides all the comfort you would get from a meat-based ragu, minus the heaviness on your digestion and will in fact support your gut because the nourishing red lentil base provides a big dose of gut healthy fibre, which encourages regular bowel movements and the growth of beneficial bacteria. Lentils are also a good plant-based protein to sustain you and balance out the carbs from the pasta and, to top it off, include a decent-dose of hormone healthy nutrients to support energy production and replenish depleted stores including B vitamins, iron, magnesium, potassium and zinc.

Rosemary really shines through in this ragu sauce, a herb known for its ability to stimulate circulation and support blood stagnation, which is particularly helpful in cases of endometriosis. In addition, it's also rich in in antioxidants to support the immune system and we don't need to highlight how wonderful it tastes!

Serves 3–4

200g mushrooms (or swap for 3 finely chopped carrots)
200g sun-dried tomatoes
1 large red onion
1 tbsp extra-virgin olive oil
15g fresh rosemary
1 tbsp dried oregano
1 tbsp dried thyme
2 tbsp balsamic vinegar
4 garlic cloves, grated
2 tbsp tomato purée
400g tin tomatoes
Bone broth powder (optional - for even more nutrients)
1 litre vegetable stock
250g dried red lentils
2 bay leaves
1 red pepper
2 servings spelt spaghetti or pasta

1. Chop the mushrooms, sun-dried tomatoes and red onion.
2. Heat the extra-virgin olive oil in a big pot and add the chopped vegetables, along with the rosemary, oregano and thyme. Stir and cook, until soft.
3. Pour in the balsamic vinegar and add the garlic, tomato purée, tinned tomatoes and bone broth powder, if using.
4. Pour in the vegetable stock and red lentils and give everything a good stir.
5. Lower the heat, bring to a simmer and pop in the bay leaves. Cover and cook for about 45 minutes over a low heat until the lentils are tender.
6. Around 10 minutes before you're ready to eat, cook the pasta according to the packet instructions (adding a pinch salt and glug of extra-virgin olive oil to the water). When it's almost ready, add a little pasta water to the lentil ragu pot (don't skip this – it makes such a difference!). Serve the cooked pasta with the lentil ragu.

TIP: Keep cut fresh herbs in a glass jar in water in the fridge to preserve them for longer and make fresh rosemary tea to sip on to support circulation related to endometriosis or for a tasty anti-inflammatory tea.

Herby Halloumi Mediterranean Veggies

If you like halloumi, you will love this. It's what halloumi dreams are made of!

During your luteal phase when you might be craving some saltiness and a meal that satisfies your taste buds as much as your gut microbes, you must give this one a go. It never disappoints! It's rich in a rainbow of roasted plants and super quick to prep. Simply serve it with quinoa or stuff into wraps for an easy, super-tasty, feelgood recipe your body will want you to make again!

Serves 2

2 servings of quinoa
2 peppers (any colour), sliced
1 courgette, sliced
1 red onion, sliced
400g tin chickpeas, drained and rinsed
Handful of cherry or vine tomatoes
1 tsp extra-virgin olive oil
250g halloumi, sliced
2 tsp Italian seasoning
1 tsp fennel seeds
1 tsp raw honey or maple syrup
Handful of rocket per bowl, to serve
Handful of olives, to serve
Small handful of basil, to serve
Balsamic vinegar, for drizzling (optional)

1. Preheat the oven to 180°C.
2. Cook the quinoa according to the packet instructions.
3. Line a roasting tray with parchment paper and place the vegetables on to the tray, along with the chickpeas and tomatoes. Drizzle with extra-virgin olive oil.
4. Place the halloumi slices in the middle of the tray and sprinkle with the herbs and fennel seeds. Finish with a drizzle of honey or maple over the halloumi.
5. Roast in the oven for approximately 25–30 minutes, or until the halloumi is golden and cooked to perfection
6. Divide the quinoa between two bowls, along with the rocket and add the roasted halloumi and vegetables on top. Finish with olives and basil, plus a little drizzle of balsamic vinegar, if you fancy it.

SPOTLIGHT:
Halloumi is made from dairy that contains a protein called A2 casein, which tends to be easier to digest and more easily tolerated than cow's dairy which contains A1 casein. A2 dairy comes from certain breeds of cow including Jersey cows, as well as dairy that comes from sheep and goats – in addition to halloumi this includes cheeses such as goats' cheese, feta, Manchego, pecorino, ricotta and Roquefort.

Gut-glow Chana Masala

Your gut microbes will love this one. It is made up of a sweet-potato base, containing two types of fibre (soluble and insoluble), which can be fermented by the bacteria in your colon, creating short-chain fatty acids (see page 67) that fuel the cells of your intestinal lining and keep them healthy and strong.

Cooking this recipe in a little fat, such as coconut oil or ghee, can help to boost the absorption of beta carotene since it's a fat-soluble nutrient.

If you can get your hands on purple sweet potatoes, that's even better! Studies have found that antioxidants in purple sweet potatoes promote the growth of healthy gut bacteria, including certain *Bifidobacterium* and *Lactobacillus* species.[66]

This extra special recipe will will also top up your nutrient stores and warm you up from the inside out during the autumn and winter phases of both your cycle and the year.

Serves 4

2 sweet potatoes
1 tbsp extra-virgin olive oil
1 tbsp ghee or extra-virgin olive oil, for cooking
2 tbsp medium curry powder
Generous pinch each of sea salt and freshly ground black pepper
5-cm piece of fresh ginger, grated
4 garlic cloves, grated
2 red onions, finely sliced
1 tbsp ground cumin
1 fresh green chilli (optional)
400g fresh tomatoes, chopped (or tinned tomatoes)
2 tbsp garam masala
1 tbsp mild chilli powder
½ tsp ground turmeric
½ tsp ground coriander
2 x 400g tins chickpeas, drained and rinsed
1 vegetable stock cube mixed into 300ml boiling water
Juice of ½ lemon
Big handful of kale or spinach (roughly 80–100g)
Big handful of fresh coriander (roughly 30g)
2 tbsp peanut butter

1. Preheat the oven to 190°C.
2. Chop the sweet potatoes into small chunks, transfer to a roasting tray, toss with the extra-virgin olive oil, 1 tablespoon of the curry powder and the sea salt and pepper and roast for 30 minutes, turning halfway through.
3. Now heat the extra-virgin olive oil or ghee in a large pot over a medium heat and add the grated ginger and garlic, chopped onions, green chilli (if using) and ground cumin and the remaining curry powder.
4. Once the onions have softened, pour in the tomatoes and all dried spices.
5. Add the chickpeas and pour in the vegetable stock water, bring to the boil, then simmer for 10 minutes.
6. Transfer the roasted sweet potato chunks to the pot, along with the lemon juice, kale, coriander (reserving a little for serving) and finally the peanut butter.
7. Give it a good stir, then serve on its own or with brown rice/quinoa (I like to mix both grains in this recipe), garnished with the remaining coriander.

Celeriac and Carrot Chopped Salad with Coconut Lemon Dressing

Celeriac is often forgotten, but it's a hearty, substantial root vegetable with impressive health benefits, including being high in fibre and vitamins B6, C and K. It's also a good source of antioxidants and important minerals, such as phosphorus, potassium and manganese.

Don't be put off by its rough exterior, as it's easy to remove and in this salad. I use it raw, alongside the humble carrot, which contains a unique fibre called lipopolysaccharides (LPS) that effectively binds to excess oestrogen and helps the body to safely eliminate it via the bowels, helping to reduce those pesky PMS symptoms.[67]

Serves 3–4

220g quinoa
½ celeriac
½ red cabbage, finely chopped
2 carrots, grated
½ red onion, finely chopped
80g almonds, roughly chopped
15g flat-leaf parsley, chopped

For the coconut lemon dressing
300g coconut yoghurt
Juice of 1½ lemons
1 tsp grated lemon zest
Generous pinch each of sea salt and freshly ground black pepper
2 garlic cloves, grated
3 tbsp extra-virgin olive oil
15g flat-leaf parsley, chopped

1. Cook the quinoa according to the packet instructions.
2. Remove the outer rough skin of the celeriac and then slice into skinny little matchsticks (don't worry about them being perfect, just as thin as you can get them). Or, for ease, use a food processor.
3. Make the dressing by mixing all the ingredients together.
4. Add all the prepared vegetables and the almonds to the biggest bowl you own, along with the cooked quinoa and parsley and drizzle with the dressing. Give everything a good mix and serve on its own or alongside your choice of protein.

TIP: Basil and coriander also work well.

'Bring-you-back-to-life' Comforting Cauliflower Curry

I created this one-pot winter warmer to literally bring you back to life during the late autumn of your cycle, when oestrogen is dipping and your body is craving wholesome, hearty food, while needing all the nutrients it can soak up and store, ready for the pending coldness of winter (your period).

High-fibre sweet-tasting root vegetables and anti-inflammatory, aromatic herbs and spices will support the large intestine and bowel motility, which is important because the rise in progesterone in the days leading up to your period can slow transit time and cause constipation for some women.[68, 69]

Serves 4

1 butternut squash
1 tbsp extra-virgin olive oil
1 tbsp ghee or extra-virgin olive oil, for cooking
2 tbsp medium curry powder
1 large red onion, chopped
3 garlic cloves
2 x 400ml tins coconut milk
1 vegetable stock cube, stirred into 40ml water
1 large cauliflower, cut into florets
400g tin chickpeas, drained and rinsed
1 tsp ground turmeric
1 tsp ground cumin
½ tsp cayenne pepper
1 tsp mild chilli powder
1 tsp ground cinnamon (or 1 cinnamon stick)
30g fresh coriander
Juice of 1 lime
250g fresh spinach
Thumb-sized piece of fresh ginger
Sea salt and freshly ground black pepper

1. Preheat the oven to 200°C.
2. Peel and chop the squash into chunks, then place on a baking tray lined with parchment paper. Drizzle with extra-virgin olive oil and season with 1 tablespoon of the curry powder and a pinch of sea salt and pepper. Roast for 35 minutes or until slightly browned around the edges.
3. Heat the ghee or oil in a big pot over a medium heat. Sauté the onion, then grate the garlic straight into the pot and pour in the coconut milk and stock.
4. Add the roasted squash and cauliflower florets, along with the chickpeas, all the ground seasoning, the remaining curry powder, fresh coriander, lime juice, spinach and ginger. Stir and then allow to simmer over a low heat for around 30–40 minutes.
5. Serve on its own or with basmati brown rice or my flatbreads (see page 200).

Tuna and Chickpea Chopped Salad

The follicular phase of your cycle represents the spring in Chinese medicine, and I've created the perfect lunch recipe to support your body during this time.

My tuna and chickpea salad combines the light and fresh salad ingredients you will be likely craving as your hormones start to rise after the cold winter of your cycle, combined with satisfying wholegrain fibre from the naturally gluten-free grains for gut and neurotransmitter support, complete protein (from the tuna) to support steady blood-sugar levels, healthy fats (from the avocado), which will hydrate your skin from within and the most divine detoxifying parsley and garlic dressing to support your detox. If tuna is your thing, this salad ticks all the boxes!

Serves 2

400g tin chickpeas
¼ red onion, finely chopped
1 avocado
100g cherry tomatoes
½ cucumber
Handful of flat-leaf parsley
Handful of walnuts
3 tbsp sliced olives
150g cooked quinoa, buckwheat or brown rice
120g tin tuna

For the dressing
1 garlic clove, grated
15g chopped flat-leaf parsley,
1 tbsp apple cider vinegar
1 tbsp raw honey
2 tbsp extra-virgin olive oil
1 tbsp lemon juice
1 tbsp grated lemon zest
Pinch of sea salt and freshly ground black pepper

1. Drain the chickpeas and pat them dry
2. Chop the remaining salad ingredients into small chunks.
3. Throw everything into a big bowl, along with the cooked grains and tuna.
4. Combine all the dressing ingredients, then drizzle over the salad and toss together. Transfer to serving bowls and enjoy.

Harissa, Butter Bean and Red Pepper Orzo

This recipe gives pasta vibes and satisfaction while also being the ultimate comfort food, with all the nutrient-dense extras to enjoy during the luteal phase of your cycle when your body is using more energy and needs extra carbs to help carry out its job in prepping for your period. It's high in fibre from the beans, rich in the antioxidant lycopene from the tomatoes (known for its potent anti-inflammatory properties and for keeping free-radical levels in balance and protecting your body against chronic health conditions such as cancer, diabetes, heart disease and Alzheimer's).[70]

The taste of the micronutrient-powered fresh basil and parsley along with the zesty freshness of the alkalising lemon contrast with the smoky-sweet harissa and just works so well. And for a complete-protein boost, add prawns, chicken or tofu. Your next one-pot dinner!

Serves 2 adults and 1 mini

Drizzle of extra-virgin olive oil, for cooking
1 red onion, finely chopped
2 medium red peppers, finely chopped
5 garlic cloves, grated
400g tin chopped tomatoes
1½ tbsp tomato purée
1 tbsp harissa paste
½ tsp dried chilli flakes (skip this if making for kids)
Pinch each of sea salt and freshly ground black pepper
500ml veg stock
150g dried orzo
400g tin butter beans, drained and rinsed
Large handful of chopped basil and flat-leaf parsley (roughly 15g of each)
Big handful of cavolo nero
1 tsp grated lemon zest
Feta cheese, to serve (optional)

1. Heat a drizzle of oil in a big pot over a medium heat. Add the onions, peppers, garlic and tomatoes and fry for a few minutes.
2. Add the tomato purée, harissa paste, chilli flakes (if using), sea salt and black pepper and stir well.
3. Pour in the stock, along with the orzo and butter beans. Add the basil, parsley, cavolo nero and lemon zest and give it all a big stir.
4. Bring to the boil, then reduce the heat and simmer for 10 minutes, stirring to prevent the orzo from sticking. Serve as it is or with crumbled feta on top and enjoy!

'PMS-free' Winter Squash Salad

If you suffer with PMS and are craving salad but you know it's not quite going to cut it (luteal-phase needs!) treat your body to my delicious cycle supportive winter salad, it's rich in cruciferous vegetables, which are especially beneficial for estrogen excess because they're naturally rich in a compound called diindolylmethane (aka DIM). This means they help break down, metabolize and eliminate estrogen after it's done its job so that it doesn't recirculate and cause hormonal imbalances driving PMS.

This nourishing salad is also high in antioxidants and phytoestrogens to promote hormonal balance meaning fewer PMS symptoms and a happier hormonal environment.

Serves 2

1 butternut squash
Extra-virgin olive oil, for cooking
Generous pinch each of sea salt, freshly ground black pepper and dried chilli flakes
1 courgette
150g frozen organic edamame beans
150g frozen sweetcorn
200g kale or cavolo nero
220g tenderstem broccoli
3 garlic cloves, grated
2 tbsp tamari
250g quinoa (I use precooked for ease)
1 tbsp white miso paste
Fresh coriander and mixed seeds, to serve

1. Preheat the oven to 190°C.
2. Peel the squash and chop into chunks. Place in a roasting tin, drizzle with 1 teaspoon extra-virgin olive oil, sea salt, pepper and chilli flakes. Toss together and then pop in the oven to roast for 30 minutes.
3. Chop the courgette and pour boiling water over the edamame and sweetcorn in a bowl and set aside.
4. Heat 1 tablespoon of extra-virgin olive oil in a big pan over a medium heat and stir-fry the courgette, kale or cavolo nero and broccoli, along with the garlic and tamari.
5. Drain the water from the sweetcorn and edamame bowl and add them to the pan.
6. Remove the squash from the oven and add to the pan, mixing into the rest of the ingredients.
7. Stir in the precooked quinoa, along with the white miso paste and heat through.
8. Serve immediately, topped with coriander and mixed seeds (and save some for tomorrow!).

Roasted Garlic Za'atar Hummus

Two hero ingredients, garlic and za'atar, are blitzed here into a silky-smooth hummus packed full of gut-feeding fibre, anti-inflammatories and flavour. Dip crudités into it for a nourishing snack, add to salads or spread on wraps or sourdough.

Yes, roasting the garlic and peeling the chickpea skin *is* an extra step, but one that's well worth the effort. Trust me, roasted garlic adds a depth of flavour that unroasted garlic just cannot offer. I love adding it to salads – it really is a magic ingredient and a nourishing staple that takes food from good to great.

This hummus lasts for four to five days in the fridge.

Makes 1 bowl

1 garlic bulb, whole
4 tbsp extra-virgin olive oil
Pinch each of sea salt and freshly ground black pepper (or to taste)
400g tin chickpeas
2 tbsp tahini
Juice of ½ lemon
½ tbsp za'atar
1 ice cube
Dried chilli flakes (optional)

1. Preheat the oven to 200°C.
2. Drizzle the garlic bulb with extra-virgin olive oil, sprinkle with the sea salt and pepper and place on a baking tray.
3. Place an upturned ovenproof ramekin over the garlic bulb (like a lid) and bake in the oven for 25–30 minutes or until golden and bubbling.
4. In the meantime, drain the chickpeas reserving 2 tablespoons of the chickpea water from the tin or jar to use later. Place the chickpeas in a large bowl and cover with tap water, rubbing the chickpeas until the skins come off (this makes your hummus super smooth!). Drain the chickpeas and discard the skins.
5. Transfer the chickpeas to a food processor, along with the remaining ingredients and the reserved chickpea water.
6. When the garlic is ready, squeeze it directly from the bulb into the food processor and blitz on the highest setting until smooth. Add a splash of water to loosen the consistency, if needed and serve.

My Best-ever Dhal

This hearty and wholesome one-pot batch-cook recipe is my best dhal yet, and one of my most loved recipes shared on social media. It's big on flavour, packed with anti-inflammatory plant diversity and gives all the comfort-food vibes needed during the autumn and winter of your cycle. Because this is when your body craves warming nourishment to keep cortisol in check and help it to carry out its important job of prepping for your period in the coming days.

Serve this as it is or with brown rice, quinoa or buckwheat.

Serves 4

3 medium-sized sweet potatoes
2 medium carrots
1 big red onion
7½cm piece of fresh ginger, grated
3 garlic cloves, grated
1 tsp turmeric
1 tsp ground cumin
1 tsp garam masala
1 tsp ground cinnamon
1 tsp mild chilli powder
2 x 400ml full-fat coconut milk
400g tin tomatoes
200g red split lentils
1 tbsp medium curry powder
1 vegetable stock cube
Big handful of spinach
Big handful of coriander
Finely chopped red and green chillies or dried chilli flakes (optional)

1. Chop the sweet potatoes, carrots (both with skin on) and red onion into small chunks. Place in a large pot over a medium heat, along with the grated ginger, garlic and all the spices (except the curry powder).
2. Pour in the coconut milk, tinned tomatoes and lentils along with the curry powder and crumble in the vegetable stock cube.
3. Bring to the boil, then leave to simmer over a low heat for a good 40 minutes.
4. Stir in the spinach and coriander and then serve, garnished with the chopped chillies or chilli flakes, if using!

Creamy Coconut Chickpea Curry

Perfect, cosy nourishment to enjoy during the winter of your cycle, your menstrual phase, when your body is asking for warming, fibre-rich food to re-energise you, satisfy your senses and feed you back to feeling you again. Make a big batch of this curry and save the leftovers for lunch the next day.

Makes 2 hearty and 2 mini bowls; or 4 hearty bowls, when served with grains

1 tbsp extra-virgin olive oil, coconut oil or ghee
2 red onions, chopped
7½cm piece of fresh ginger, grated
5 garlic cloves, grated
2 tbsp medium curry powder
1 tbsp garam masala
1 tbsp ground cumin
1 tsp cumin seeds
1 tsp mild chilli powder
Pinch of cayenne pepper
Big twist of black pepper
2 x 400ml tins coconut milk
1 vegetable stock cube
2 tbsp maple syrup
3 tbsp peanut butter
2 x 400g tins chickpeas, drained and rinsed
400g tin black beans, drained and rinsed
1 cinnamon stick (or 1 tsp ground cinnamon)
200–300g spinach
Big handful of fresh coriander (approximately 30g)

1. Heat the extra-virgin olive oil or ghee in a big pot over a medium heat before adding the onion, ginger and garlic. Sauté for a few minutes until soft.
2. Now add all the ground spices and seasoning, pour in the coconut milk and crumble in the vegetable stock cube before adding the maple syrup and peanut butter.
3. Stir in the chickpeas and black beans, then add the cinnamon stick and leave to simmer over a low heat for approximately 30 minutes.
4. Remove the cinnamon stick, add the spinach and coriander and give it a good stir. Let your curry simmer for a final 5 minutes before serving on its own or with brown rice.

TIP: You can adjust quantities here to your own taste preferences. For example, if you like it creamier, add more peanut butter, or to make it more fiery, use more ginger and chilli powder (or chilli flakes).

Supercharged Sweet Potato and Smoky Beans

Ideal to eat during the autumn and winter of your cycle because it's warming and nourishing, my levelled-up version of the humble jacket potato and baked beans has so much more nutrition to offer your hormones than the old-school classic. Rich in fibre and abundant in antioxidants, sweet potatoes are a prime example of nature's carbs to help energise your body when it needs it. Some research suggests that this versatile root vegetable naturally encourages progesterone production (and studies show that it's a fantastic source of vitamin A, linked to helping healthy thyroid function).[71]

Black beans are not only a plant-based (non-haem) iron to increase stores in the lead up to and during your period, they're also a good plant protein, helping to balance out the complex carbs in the sweet potato and, therefore, constituting a restorative meal with dynamic flavours to please not just your hormones but your tastebuds, too.

Serves 2

2 medium-sized sweet potatoes
1 tbsp extra-virgin olive oil, plus extra for drizzling
1 spring onion, finely chopped
1 x 400g tin black beans, drained and rinsed (or mixed beans)
1 tbsp smoked paprika
½ tsp ground cumin
½ tsp mild chilli powder
½ tsp ground cinnamon
1 red pepper, finely chopped
2 garlic cloves, grated
2 tbsp tomato purée
1 tbsp maple syrup or raw honey
100ml water (plus more to loosen, if needed)
15g fresh coriander, roughly chopped
Pinch of dried chilli flakes, to taste
Sea salt and freshly ground black pepper
Slices of avocado or guacamole (see page 59), to serve

1. Preheat the oven to 190°C.
2. Score the sweet potatoes, drizzle with extra-virgin olive oil and season with sea salt and pepper. Pop in the oven for 30 minutes or until cooked to your liking.
3. In the meantime, heat the extra-virgin olive oil in a pan over a medium heat and add the spring onion, beans, ground spices, chopped red pepper, garlic, tomato purée, maple syrup, water, sea salt and pepper.
4. Leave to simmer and once it starts to thicken, reduce the heat to low and leave for approximately 15 minutes, stir the chopped coriander through before serving
5. Slice open the sweet potatoes and fill with the smoky bean mix, top with chilli flakes, slices of avocado or guacamole and dive in!

CYCLE SYNC | ALL PHASES FOR GUT-HEALTH SUPPORT; FOR CYCLE SUPPORT SECOND PART OF LUTEAL AND DURING MENSTRUAL

Sunday-night Nourishing Chicken Bone Broth

Bone broth is truly one of the most nourishing recipes you can support your hormones with and is one of my favourite post-Sunday roast rituals. I love knowing that every single part of the roast has been maximised for zero wastage and maximum hormone-health benefits.

Containing a powerhouse of nutrients, including collagen, amino acids, magnesium and phosphorus, this anti-inflammatory, healing elixir will support your system, from encouraging healthy hair and nails to soothing your gut microbiome and promoting skin elasticity and it is also incredibly supportive of fertility (I drank it daily during my own journey).

Serves 4

1 roast chicken carcass (see recipe, page 120)
2 medium carrots, unpeeled and chopped
2 onions, chopped
2 celery sticks, chopped
4 garlic cloves, grated
1 tsp sea salt
1 tsp freshly ground black pepper
2 tbsp organic apple cider vinegar
A few sprigs of fresh herbs (I like to use rosemary or thyme)

1. Place the chicken carcass in a big pot and cover with filtered water.
2. Add the carrots, onions, celery, garlic, sea salt, pepper, vinegar and fresh herbs.
3. Bring to the boil and then simmer over a low heat for at least 3 hours (or longer for a more intense flavour).
4. Strain the bone broth into a big bowl and then either add the meat from the bones back in (I like to do this) or discard.
5. Store in a glass, airtight container in the fridge for a maximum of 4 days or pour into ice-cube trays to add to sauces.

TIP: Try it on its own to lift your day or pour into ice-cube trays and add to any soup or sauce-based recipes, from pasta sauces to curries and stews, for a big boost of goodness.

Healing 'Hug-in-a-Bowl' Chicken Soup

When you're in need of a hug-in-a-bowl of goodness, look no further! This healing, super-nourishing recipe is the sort of food my granny used to love, and I always think of her when I make it. It couldn't be more loaded with health-enhancing anti-inflammatory compounds, proven to reduce inflammation and encourage the body's healing process. Truly a recipe that makes you feel held and taken care of.

I make this during my period when my hormones are plummeting and I need a nutritional boost, or when I'm feeling run down and my immune system could do with a little help.

Serves 4

1 tbsp extra-virgin olive oil
2 medium carrots, chopped very small
3 celery sticks, cut into small chunks
1 big brown onion, cut into small chunks
6cm piece of fresh ginger, grated
1 tbsp fresh grated turmeric root (or 1 tsp ground turmeric)
3 garlic cloves, grated
1 tbsp rosemary
1 tbsp thyme
Generous pinch of cayenne pepper
Pinch each of sea salt and freshly ground black pepper (or to taste)
1.3 litres chicken or vegetable stock or broth
250g red lentils
1 bay leaf
150g frozen peas
1 tbsp raw honey or maple syrup
300–400g shredded, cooked chicken (I use roast-dinner leftovers – see my Go-to Garlic-rubbed Roast Chicken, page 120) or poached and shredded chicken (see method opposite)

1. Heat the extra-virgin olive oil in a big pot over a medium heat and sauté the carrots, celery and onion until soft.
2. Add the ginger, turmeric and garlic to the pot, followed by the rosemary, thyme, cayenne pepper and sea salt and pepper.
3. Pour in the stock followed by the red lentils. Give it a good stir and pop in the bay leaf. Cover and leave to simmer over a low heat for about 15 minutes.
4. Add the frozen peas, maple syrup and cooked chicken and leave to simmer for another 5 minutes to heat the chicken through before serving. Enjoy the healing nourishment!

HOW TO POACH AND SHRED CHICKEN

1. Add boneless skinless chicken breasts to a large pot of water and bring to the boil.
2. Boil for 10 minutes or until chicken is no longer pink.
3. Remove from pot and allow to cool for a few minutes.
4. Use two forks to shred and pull the chicken apart into smaller pieces.

Nutty, Seedy, Slightly Spiced Granola

Making a batch of homemade granola is such a wholesome ritual and always sets me up for the week ahead; then, when breakfast rolls around, I know exactly what's in it, and that there are zero hidden nasties. This granola is the perfect way to consume multiple plant varieties and, therefore, gut-loving fibre, vitamins and minerals, plus it adds a satisfying, crunchy joy to your breakfast. Or enjoy whenever you need a healthy, gut-loving snack.

I recommend serving this with live, cultured yoghurt (such as organic, unflavoured and unsweetened Greek yoghurt) and adding juicy berries for the antioxidants and sweetness. I'm a big fan of authentic Greek yoghurt because it contains a host of beneficial live bacteria that help to maintain a thriving microbiome and healthy digestive system. It's also a rich source of calcium and a complete protein, meaning it contains all nine essential amino acids for optimal body functioning.

Serves 6

200g gluten-free jumbo oats
100g buckwheat
120g mixed raw nuts (I use a mix of almonds, walnuts and pecans), roughly chopped
100g mixed seeds (I use hemp, chia, pumpkin and sunflower or pumpkin depending on the cycle phase I'm in)
2 tbsp melted coconut oil or MCT oil
4 tbsp raw honey or maple syrup
1 tsp ground cinnamon
½ tsp ground ginger
3 tbsp nut butter (peanut, almond or cashew)
Generous pinch of sea salt

1. Preheat the oven to 160°C.
2. Mix the oats, buckwheat, nuts and seeds in a big bowl. Add the melted coconut oil, honey, ground cinnamon, ground ginger, nut butter and sea salt. Mix well, so everything is coated nicely.
3. Transfer the granola mix onto a large baking tray lined with parchment paper and spread it out evenly, pressing and firming down with your fingers. If you like your granola in clusters or clumps, then again, using your fingers, gently press some of the mixture together to create small chunks.
4. Roast in the oven for 30 minutes, mixing it halfway through, so that it's evenly toasted and doesn't burn.
5. Remove from the oven and leave to cool properly before pouring into a glass jar or airtight container. It will keep well for a few weeks.

SPOTLIGHT:
This recipe includes buckwheat for the added fibre, antioxidants, magnesium, B vitamins and plant protein.

Roasted Cauliflower, Cashew and Buckwheat Bowl with Tahini Dressing

Meals that make humble vegetables like cauliflower the main event, instead of a bonus on the side are my kind of meals. And this is the perfect example! The high fibre content of the cauliflower will keep your bowels regular and the tahini dressing, rich in B vitamins, will support your energy levels during your luteal and menstrual phases, as well as being a good source of copper, a trace mineral that's essential for iron absorption. All in all, a winning luteal-phase energy-supporting combo!

Serves 2

1 cauliflower
400g chickpeas, drained and rinsed
Handful of cashews
Handful of dried cranberries
2 servings of buckwheat
2 big handfuls of watercress
15g mixed mint and flat-leaf parsley
2 spring onions, finely chopped
½ pepper (any colour)

For the harissa marinade
1 tbsp harissa
1 garlic clove, grated
3 tbsp extra-virgin olive oil
Juice of ½ lemon
Pinch of sea salt and freshly ground black pepper

For my gut-loving tahini dressing
2 tbsp tahini
2 tbsp toasted sesame oil
½ tsp za'atar
2 tsp Dijon mustard
2 tbsp maple syrup or raw honey (add to your liking)
1 garlic clove, grated
Juice of ½ lemon
Pinch each of sea salt and freshly ground black pepper
3–4 tbsp ice-cold water to loosen consistency
Pinch of dried chilli flakes (optional)

1. Preheat the oven to 200°C.
2. Rinse and dry the cauliflower. Chop into smallish chunks and add to a roasting tray, along with the chickpeas.
3. Combine all the ingredients for the harissa marinade, then pour over the cauliflower and chickpeas to coat.
4. Pop the tray into the oven and roast for approximately 35 minutes. Halfway through cooking, add the cashews and cranberries and toss it all together.
5. Prepare the tahini dressing by combining all the ingredients and slowly adding the ice-cold water, until you've reached your desired consistency (I like it on the thicker, creamier side). Season to taste.
6. Cook the buckwheat according to packet instructions.
7. Finally, when the cauliflower and chickpeas are ready, assemble your bowl by adding the watercress, herbs, spring onions, pepper, and buckwheat. Add the roasted cauliflower and chickpeas and drizzle with the tahini dressing over the top.

Speedy Italian Butter-bean Stew

Hormone-healthy, hearty, rich in gut-loving plant diversity and also super easy to throw together, my Italian butter-bean stew is a healthy hormone staple. A tasty and faff-free, one-pot batch cook that will feed your gut microbes with fibre and leave you feeling nourished.

Serve cold during follicular phases and hot during luteal phases. It tastes great served on its own or with quinoa for extra plant protein and wholegrain fibre, to keep you fuller for longer. When I make this for dinner, I serve it with a side of roasted white fish and it absolutely makes it.

Serves 2

1 tbsp extra-virgin olive oil
2 red peppers, chopped into medium-sized chunks
1 large red onion, chopped into medium-sized chunks
4 garlic cloves, grated
1 tbsp tomato purée
2 tbsp Italian seasoning
1 tsp fennel seeds
Pinch each of dried chilli flakes, sea salt and freshly ground black pepper
400g tin tomatoes
140g sun-dried tomatoes
1 tbsp balsamic vinegar
2 x 400g tins butter beans, drained and rinsed
200ml vegetable stock
250g quinoa
15g flat-leaf parsley
15g basil
200g cavolo nero, chopped

1. Heat the oil in a big pot over a medium heat. Add the peppers, onion and garlic, then stir before adding the tomato purée and all the dry seasoning.
2. Pour in the tinned tomatoes, sun-dried tomatoes, balsamic vinegar, butter beans and stock and leave to simmer over a low heat for 15 minutes.
3. In the meantime, cook the quinoa according to the packet instructions.
4. Chop the parsley and basil and add to the pot, along with the cavolo nero. Stir and leave to cook for a final 5 minutes before serving in a bowl with the cooked quinoa.

Honeyed Halloumi and Borlotti Bean Salad

Grilled honeyed halloumi, roasted tomatoes (because roasting not only brings out their sweet flavour, but research shows it also increases the bioavailability of antioxidant lycopene, known for its anti-inflammatory and anti-cancer properties), a base of PMS-soothing, magnesium-rich leafy greens and gut-health borlotti beans – what's not to love?[72,73]

Serves 2

6 slices of halloumi
Big handful of cherry or vine tomatoes
200g cooked quinoa, buckwheat or mixed wholegrains
2 big handfuls of mixed spinach and rocket
400g tin borlotti beans, drained and rinsed
1 yellow pepper, sliced
½ red pepper, sliced
¼ red onion, finely sliced
100g olives
Half a handful each of flat-leaf parsley and mint, roughly chopped

For the halloumi seasoning

2 tbsp raw honey
2 tsp rosemary, oregano or za'atar seasoning
Pinch of freshly ground black pepper

For the salad dressing

4 tbsp extra-virgin olive oil
2 tbsp apple cider vinegar
1 tsp Dijon mustard
1 tbsp maple syrup
Pinch of the same dried herbs you used for the halloumi

1. Preheat the oven to 190°C.
2. Drizzle the halloumi slices with the halloumi seasoning ingredients. Transfer to a roasting tray, along with the tomatoes and roast for around 10–15 minutes, until golden.
3. Combine the salad-dressing ingredients.
4. Divide the leafy greens and borlotti beans between two bowls, add the peppers, red onion, olives and fresh herbs.
5. Remove the halloumi and tomatoes from the oven and add to the bowls, then drizzle the dressing over the top.

TIP: Adding beans to a salad is an easy way to amplify it with fibre to feed your gut microbes. This is because as your microbes ferment fibre from the beans they produce short-chain fatty acids (SCFAs), which are vital for supporting your intestinal epithelial cells (the cells that line your gut) and keeping the cells tightly linked together, preventing 'leaky gut' (see page 67). SCFAs also play an important role in helping you to digest your food and absorb nutrients, as well as protecting against infections. Basically, eat more beans – because this convenient and cost-effective addition will keep you fuller for longer and will promote a healthy, well-nourished gut garden!

Sweet Potato Salad with Medicinal Ginger Dressing

You can't beat a hearty high-fibre salad, especially when it tastes as good as this one and has endless benefits to offer your microbiome.

Its main gut-glow benefits come from the two key ingredients – puy lentils and sweet potato, both of which are good sources of soluble-fibre, which means they draw water into your gut, and this softens your stools and supports regular bowel movements. Soluble fibre also helps you to feel fuller and reduces constipation, and it may also lower your cholesterol and blood-sugar levels.[74] These guys are also the perfect food source for good bacteria in your large intestine.

It wouldn't be a gut-glow salad without curly kale, a hormone-balancing and liver-detoxifying vegetable that's rich in sulphur compounds to keep excess oestrogen at bay (along with its pesky symptoms). To top it off, it's loaded with an array of antioxidants, anti-inflammatories and anti-fungal properties for a healthy gut from the gingery gut-glow dressing. Plus, it's a lunch batch recipe and the dressing keeps well in a glass, airtight container in the fridge for three days.

Serves 2

2 small sweet potatoes
1–2 tbsp extra-virgin olive oil
½ tbsp ground cumin
Pinch each of sea salt and freshly ground black pepper
Big handful of kale
200g puy lentils
½ Romano pepper, chopped
Handful of cherry or vine tomatoes
2 spring onions, finely sliced
2 tbsp crumbled feta
2 tbsp pomegranate seeds
1 tbsp crushed walnuts
Handful of fresh coriander, roughly chopped

For the medicinal ginger and honey dressing

4 tbsp extra-virgin olive oil
2½cm piece of fresh ginger, grated
2 tbsp apple cider vinegar
2 tsp raw honey
1 tsp Dijon mustard
Pinch each of sea salt and freshly ground black pepper

1. Preheat the oven to 190°C.
2. Chop the sweet potatoes (leave the skin on) into small chunks and drizzle with ½–1 tablespoon extra-virgin olive oil, season with the ground cumin, sea salt and pepper and roast for 35 minutes, or until perfectly golden and slightly browned around the edges.
3. In the meantime, combine the ingredients for the Medicinal Ginger and Honey Dressing.
4. Now it's time to construct your salad, adding the kale, spooning in the lentils, pepper, tomatoes, spring onions and the roasted sweet potatoes.
5. Drizzle with the delicious ginger dressing. Finally, sprinkle the feta, pomegranate seeds, crushed walnuts and coriander over the top and enjoy.

Seasonal Prebiotic Traybake with Tahini Dressing

Prebiotic vegetables such as artichoke, asparagus, leek, mushrooms and parsnips are quite literally food for the good bacteria in your gut, keeping your microbiome well fed, so it can flourish.

Prebiotics optimise the health of your gut which, in turn, supports your overall health and wellbeing, including your hormone balance (because the gut metabolises used hormones and is essential for the elimination of toxins and chemicals that can otherwise lead to hormone imbalances). A happy gut really is the key to healthy, balanced hormones.

Serves 3, if pairing with grains and greens

250g Jerusalem artichokes
200g parsnips
200g mushrooms
200g leeks
1 onion (brown or red), peeled
½ garlic bulb
2 tsp extra-virgin olive oil
½ tbsp maple syrup or raw honey
Generous pinch of sea salt (I use garlic sea salt) and black pepper
Pinch of dried chilli flakes (optional)
150g asparagus

To serve
Cooked grains
Handful of almond or cashews
Handful of leafy greens (spinach and rocket), to serve
Gut-loving Tahini Dressing (see page 98)

1. Preheat the oven to 200°C.
2. Wash, dry and chop all the vegetables (leaving the skin on) and place them (except the asparagus) in a roasting tin. (Avoid overcrowding the tray, otherwise they won't crisp up.) Drizzle with extra-virgin olive oil, ½ tbsp maple syrup or honey to help them caramelise even more, and a pinch of sea salt and pepper (plus the chilli flakes, if using). Roast for 20 minutes.
3. Now add asparagus to the roasting tin and pop back into the oven for approximately 15 minutes, until nicely browned
4. Serve the vegetables in a bowl with some grains, the almonds or cashews, a handful of leafy greens and drizzle with the tahini dressing.

Sweet Potato and Chickpea 'Happy-gut' Wraps

We can't get enough of these wraps in our house – they're a real crowd-pleaser! I refer to them as 'happy-gut' wraps because they pack an impressive punch in terms of gut-feeding fibre, keeping our guts happy (aka regular), and, as a bonus, they are undeniably tasty and satisfying. The Middle Eastern spices from the za'atar, along with the sweetness of the blood-sugar-balancing cinnamon is the perfect combo. And topped with juicy pomegranate and fresh herbs, these wraps really tick all the boxes, including being great for sharing.

Serves 2

2 sweet potatoes
400g tin chickpeas (or butter beans), rinsed and drained
1 tbsp extra-virgin olive oil
1 tbsp za'atar
1 tbsp dried rosemary
½ tsp ground cinnamon
Pinch of dried chilli flakes (or to taste)
Generous pinch each of sea salt and freshly ground black pepper
2 wraps or flatbreads (see page 200)
Hummus, for spreading
2 spring onions
2 tbsp pomegranate seeds, to serve
Handful of mint, flat-leaf parsley or coriander, to serve

1. Preheat the oven to 190°C.
2. Chop the sweet potatoes into small chunks and pat dry the chickpeas (the drier they are, the more they will crisp up). Place on a lined roasting tin, drizzle with the extra-virgin olive oil and season with all the dried seasoning, salt and pepper, then roast in the oven for approximately 40 minutes, or until golden.
3. Warm your wraps (or make your flatbreads) and spread with a little hummus. Top with the roasted sweet potato and chickpea (or bean) mix, add some of the chopped spring onions, pomegranate seeds and herbs, then fold up and enjoy.

Cajun Tuna Steak Beany Bowl

High in both animal and plant protein for steady blood-sugar levels, meaning fewer energy dips and maximised amino-acid profile, this meal takes nourishment to another level. And the best bit? It's quick to prep and is so incredibly supportive of your gut health and, therefore, hormone metabolism and, ultimately, hormone balance.

The beans are a good source of zinc, calcium, manganese, potassium and selenium, as well as folate and B vitamins, which are all essential hormone-nourishing nutrients. They are, in my opinion, so underrated, and will forever be a go-to hormone health superfood base for many of my meals. Not only are they cost-effective, they are also super satisfying and contain a powerhouse of phytonutrients (including flavonoids and anthocyanins) and, what's more, they're rich in polyphenols (the antioxidants that feed your gut bacteria). This is important because when your gut bacteria are fed their favourite type of food (fibre), they metabolise it producing the beneficial compounds I've mentioned throughout this chapter (yes, those superhero short-chain fatty acids – SCFAs), which do many amazing things for your overall health and wellbeing, one being to stimulate the development of intestinal wall cells. This helps to increase the surface area of your gut, and a larger surface means more effective mineral absorption, so you extract more goodness from your food.

Serves 2

1 lime
1 tsp Cajun seasoning
Pinch of sea salt and freshly ground black pepper
2 tuna steaks
400g tin borlotti beans, drained and rinsed
400g tin butter beans, drained and rinsed
Handful of cherry or vine tomatoes
½ red onion
15g mint
2 ripe peaches
2–3 tbsp extra-virgin olive oil

1. Slice the lime in half and squeeze into a shallow dish along with the Cajun seasoning, sea salt and pepper. Add the tuna steaks and turn until well coated. Cover and refrigerate for 1 to 2 hours.
2. Pat the beans dry and finely slice the tomatoes, onion and mint. Slice the peaches into slightly larger chunks and transfer everything into a big bowl. Drizzle with about 1 tablespoon extra-virgin olive oil.
3. Heat the remaining oil in a pan over a medium heat and pan-fry the tuna steaks for approximately 3–4 minutes on each side, until cooked.
4. Divide the peach mixture between two bowls and place the tuna steaks on top.

TIP: Feel free to swap the beans for any you have in your store cupboard and switch tuna with any other protein source, including tofu or tempeh. The taste is equally great.

5. The Liver

'Look after your liver and it will look after you.'

If you want to balance your hormones and optimise your health, there's no two ways about it – you absolutely must take the health of your liver seriously and show this vital organ the love and attention it needs. Because it's impossible to balance your hormones without also optimising the health of your liver, and I'm going to show you how to start doing this via liver-loving foods and simple lifestyle hacks. But first, here's a little liver 101.

What Does Your Liver Do?

Your liver is a powerhouse. It's truly one of the hardest-working organs in your body and a real pro at multi-tasking so many of your body's major processes. It's responsible for keeping your blood clean, your metabolism functioning well, your digestive system strong and is the CEO of regulating your hormones.

Your liver is your internal filter system. Everything you swallow, breathe in and even absorb through your skin is filtered through your liver. And the most impressive part is that it does this naturally and automatically by separating the nutrients needed for energy and health and disposing of the many substances, such as toxins (from alcohol, refined sugars, pollution) and metabolic waste your body doesn't need and won't benefit from. Clever, right? But what happens when it starts to slow down?

SLUGGISH LIVER = SLUGGISH BODY (AND MIND)

A sluggish liver is a warning sign that your body isn't flushing out/detoxifying toxins properly and means your liver is struggling to function optimally. Not ideal at all. You might be able to determine if yours is on the sluggish side if you are suffering/dealing with regular skin breakouts (more on this in Chapter 6), low energy, fatigue, constipation, weight gain and low mood, among other undesirable symptoms.

This is a problem and one you need to fix, with the help of this chapter – because aside from disrupting your hormones, a slow liver can prevent other key systems (including digestive, immune and kidneys) from working at their best, too, creating a sluggish internal cascade.

The Liver–Hormone Connection[75]

Your liver is in charge of metabolising your hormones, including thyroid (both activating and inactivating them), glucagon-like peptide-1 and steroid hormones.[76,77,78] This is a fancy term that essentially refers to the liver breaking the hormones down and getting rid of them once they've served their purpose.

When your liver is compromised, you can end up with an accumulation of unwanted hormones circulating your body via your blood, which causes hormonal imbalances. A sluggish liver is incredibly common – I see it in my practice daily – but the good news is that it's also incredibly reactive to food and lifestyle interventions.

Take oestrogen as a key and common example here when it comes to a compromised liver impacting hormone metabolism. Too much of this key sex hormone circulating in your bloodstream (formally known as 'oestrogen dominance') is associated with, and can lead to, PMS symptoms, heavy periods, fatigue, weight gain, breast tenderness, water retention, fibroids, migraines and headaches, not to mention rocking the boat with her partner in crime, progesterone. And it can put you at increased risk of developing breast cancer, too.[79]

The liver also has a relationship with the hormone insulin, which is very important when it comes to your metabolic health and weight. High levels of insulin signal to your liver to store more fat, which, in serious cases, can lead to a condition called NAFLD (non-alcoholic fatty liver disease).[80] This, as the name suggests, is a build-up of fat in the liver and takes the concept of sluggish to the next level.

Endocrine Disruptors (EDCs) – and Why They Matter[81]

As we know, the liver's primary job is detoxification, and it is one hell of a role! Frankly, it doesn't get the credit it deserves, especially given the high levels of endocrine-disrupting toxins it's generally exposed to. These may come from our diets (the Western diet is high in refined sugars, pro-inflammatory chemicals and pesticide-sprayed produce that the liver then has to filter) or our environments (think car and train fumes on the commute to work, perfume you either spray on yourself or breathe in from other people, household cleaning products, candles, air fresheners, cigarette smoke – your own or second-hand – the plastic you store your food in, heat up in the microwave or drink your water out of and the beauty products you apply every morning). All of these are endocrine disruptors – natural or human-made chemicals that mimic, block or interfere with our hormones.

THE TRUTH ABOUT EDCS (AKA 'HORMONE MIMICKERS')

EDCs are synthetic chemicals or compounds that interfere with the way the body's natural hormones work and can cause problems when it comes to balancing hormones by producing adverse effects on the endocrine glands. This is because some EDCs trick our bodies into thinking that they are hormones (hence their alias), while others block our natural hormones from carrying out their jobs. Still other EDCs can increase or decrease the levels of hormones in our blood by affecting how they are made, broken down or stored in the body.

When hormones are released by your glands they travel around your body in the bloodstream, stopping when they reach their receptor points. They then bind to these points and cause a reaction. The easiest way to understand how this biochemical reaction works is to think of it as a lock-and-key system, whereby your hormones (the keys) will only act on a part of the body (the lock) if they fit.[82] The pesky endocrine disruptors create an issue for this lock-and-key mechanism, however, because they can trick the body into thinking they are naturally produced hormones when, in fact, they are just mimickers. Mind-blowing stuff, and it must not be underestimated, as these hormone mimickers commonly cause an excess in oestrogen, driving hormone imbalances and worsening related symptoms.

So you can see why EDCs have been linked to many adverse hormone-health problems, including compromised reproductive health, endometriosis, PCOS, as well as altered nervous-system functioning, metabolic issues, diabetes, cardiovascular problems, ADHD and more.

All this may feel a little overwhelming, but it's crucial to highlight the importance of reducing your toxic load to support your hormone balance and overall health and wellbeing because every one of these toxic chemicals is filtered through your liver, causing a strain on your body and its resources.[83] But please don't panic, because while we cannot control all external factors, we can take steps to significantly reduce our exposure, so helping our hormones on a daily basis.

A FEW COMMON ENDOCRINE DISRUPTORS TO BE MINDFUL OF [84]

BPA (bisphenol A): found in plastics, including water bottles, baby bottles, receipts from shops, the lining inside tins of food, dental sealants, water pipes.

Phthalates: found in food packaging, Tupperware (keep out of the microwave, dishwasher and freezer, as hot and cold temperatures cause the phthalates to be released), cling film, shampoo, hairspray, perfume and nail varnish.

Parabens: found in hair products, shower gel, soap, toothpaste, beauty products.

Pesticides: used on fruit and vegetable crops and garden products (it's also likely that non-organic meat sources have eaten pesticide-laden feed).

PFAS (per- and polyfluoroalkyl substances): a large group of chemicals used in food packaging, non-stick pans, fire-fighting foams, paints, cleaning products and drinking water (due to contamination).

Dioxins: often found in bleached products, such as tampons and sanitary towels.

Solvents: found in nail varnish and remover.

The best ways to evade hormone disruptors is to opt for organic produce where possible, avoid using plastics (water bottles, food containers, cling film), drink filtered water, swap non-stick cookware for stainless steel or cast iron and steer clear of products that contain parabens, fragrances, phthalates (if unsure, check online or use one of the many apps available). Try also to choose fragrance-free, chemical-free and eco-friendly personal-care products.

XENOESTROGENS AND HOW TO AVOID THEM

Xenoestrogens are a specific type of endocrine disruptor that exert oestrogen-like effects and are found in pretty much all plastics (and found in the oral contracpetive pill too!)[85]. The chemical structure of xenoestrogens is almost identical to that of natural oestrogen. So when xenoestrogens enter the body, they have the ability to increase oestrogen levels and bind to oestrogen-receptor sites in the same way that natural oestrogen does. This can result in a build-up of oestrogen, causing an excess (see page 29).

Reduce Toxin Exposure for a Healthier Liver and Happier Hormones

The fewer toxins you expose yourself to, whether in the foods you eat or the chemicals you use, the happier and healthier your liver will be, and the better its ability to metabolise your hormones, excrete them from your body and, therefore, maintain hormonal equilibrium.

To best support this vital liver–hormone relationship (including keeping oestrogen and, therefore, progesterone in a balanced state), my advice is to focus less on an extreme 'liver detox', as I see so often on social media, and, instead, look after your liver every single day. That's what will have the biggest positive impact on your hormone balance.

By adopting the simple swaps outlined below, you will actively help your liver to focus on carrying out its main detoxification role (and to do so with ease), so that it doesn't have to work extra hard filtering through all the toxins. Here's how . . .

EASY WAYS TO SUPPORT YOUR LIVER

Let's be real here. You can't stop the fumes from the cars you walk past or from the train you get to work on (although prioritising time spent walking in nature will help by bringing more oxygen into your cells and helping your lungs to expel more toxins from the body). But you absolutely can increase the phytochemicals and fibre in your diet, ramp up your hydration, make mindful choices when it comes to the organic produce you buy in your weekly food shop and start questioning your toxin exposure.

Start by following the liver-enhancing recipes in this chapter (see pages 119–141).

- Incorporate the liver-supportive ingredients listed on page 116 into your daily diet, adding them to your meals as seasoning and making teas from them, which will also double up as anti-inflammatory hydration – win-win!
- Prioritise organic produce where possible, but if your budget doesn't allow for this, don't worry because every year the Environmental Working Group tests all produce for the highest and lowest pesticide contamination and shares 'the dirty dozen' and 'clean fifteen' lists. This always helps me to make mindful decisions on what I should prioritise for buying organic and what's been less exposed to pesticides.
- Ensure all your fresh produce is soaked in water with a little apple cider vinegar (1-part vinegar to 4-parts water) and that the skin is brushed (using a vegetable brush) to remove as many toxins as possible before eating.
- Eat bitter greens (such as rocket, artichokes, watercress, parsley, chard, dandelion, burdock and chicory) before or at the start of your meal. This Ayurvedic practice stimulates your digestive enzymes and promotes natural liver detoxification (it's a great anti-bloat tool, too).
- Include sulphur-rich foods daily (cruciferous vegetables, broccoli – and for a super-charged effect, broccoli sprouts – kale, cauliflower, onions and garlic). Toxins bind to these foods and are then excreted out of your body.

- ✻ Eliminate alcohol if you are suffering with hormonal imbalances; it has zero benefits and taking it out will make your liver very happy indeed. Instead, infuse fresh herbs (such as rosemary and mint) in sparkling water with a big squeeze of lemon and a few grates of ginger and serve over ice in a fancy glass.
- ✻ Try my Beetroot and Parsley Liver-support Juice (see page 132) to activate your liver's enzymes, which increases bile production and helps your liver detoxify even more effectively.
- ✻ Drink 1.5–2 litres of fresh filtered water each day.
- ✻ Start your day with an immunity tea made from my handy Immunity Tea cubes (see page 170).
- ✻ Be intentional about the brands of household and beauty products you use (or replace them with when they run out).
- ✻ Avoid drinking out of plastic, as the BPA leaches into the water and therefore into your body, which your liver then must filter. Drink from glass or aluminium and avoid heating foods in plastic for the same reason – use glass containers as a non-negotiable.
- ✻ And don't forget about the vaginal microbiome and the endocrine disruptors you might be putting inside your vagina without realising. It is lined with a mucosal layer and bacterial community and anything that comes into contact with it can disrupt it. It easily absorbs chemicals through its mucus membrane, so be savvy when it comes to tampons, laundry detergent, vaginal soaps (which are totally unnecessary – our vaginas are self-cleaning) and lubricants, as these can all negatively disrupt your vaginal microbiome. Use The Environmental Working Group 'Skin Deep' Database to help find toxic-free products. As a natural replacement for lubricant, try coconut oil; it's naturally antimicrobial and won't cause the bacteria in your vagina any issues.

LIVER-LOVING FOODS TO INCORPORATE INTO YOUR MEALS[86]

- Ginger
- Garlic
- Turmeric
- Parsley
- Milk thistle
- Ginseng
- Green tea
- Liquorice root
- Beetroot
- Lemons

The Nourish Method Liver Reset Ritual

Your liver works with your lymphatic system, kidneys and digestive system in harmony to detoxify your body and maintain internal balance. The Nourish Method Liver Reset supports these systems to promote circulation and encourage toxins to leave your body for optimised detoxification, making it the perfect daily remedy and one that can be easily incorporated into your routine.

Tongue scraping: Every morning and evening before brushing your teeth, scrape your tongue. This prevents toxins being reabsorbed into the body and helps with your detoxification pathways (particularly a sluggish liver) and overall homeostasis. Use a u-shaped tongue scraper across your tongue, ideally one made of copper as it's antimicrobial.

Dry body brushing: Dry brush before showering daily to boost circulation, stimulate the lymph nodes and help your body remove waste. The goal is to brush each part of your body towards your heart in small strokes, starting at your feet and brushing upwards to your torso. When you get to your arms, start at your hands and work upwards. The chest is where the lymph system drains, so it's best to brush that area in a circular motion. Use firm pressure, taking care not to press too hard to avoid causing any pain or discomfort.

Stay hydrated: Rehydrate as soon as you wake up, as we lose hydration while we sleep. Leave a bottle of filtered water by your bed and drink it before you even get out of bed. The liver needs adequate water intake to stay efficient and carry out its detoxification job.

Legs up the wall: Stretch your legs up against a wall for twenty minutes before you go to sleep. This restorative yoga pose has been practised since the seventeenth century, and with good reason, because it supports lymphatic flow and is great for circulation, which gets things moving and helps your body detoxify.

1. Sit with your right side against the wall, with bent knees and your feet drawn in towards your hips.

2. Swing your legs up against the wall, as you turn to lie flat on your back.

3. Place your hips against the wall or slightly away from it and your arms in any comfortable position.

4. Stay in this position for up to 20 minutes –habit-stack like I do and read a chapter of your book while you do this!

I hope this chapter has alerted you to the need to address the key toxins that *are* within your control from food and lifestyle to help your liver help your hormones. And I'm going to end in the same way that I began the chapter (because once is not enough): look after your liver and it will look after you.

Liver Recipes

Go-to Garlic-rubbed Roast Chicken

I've never been a roast girl – always found them a bit mediocre. But I crave them every Sunday now and it's because this delicious recipe (which is my husband, James's!) is a 20/10 and gives me so much joy. Not just the taste, but the ritual of eating it together around the table as a family complete with our sausage dog Jeff by our feet hoping something will drop on the floor!

It's the ultimate nourishing roast-chicken recipe, rich in quality complete protein and an abundance of bioavailable nutrients to feed you and your family with.

Serve with roast potatoes and vegetables for a classic Sunday roast or with leafy greens, asparagus, and buckwheat for a spring/summer take on a weekly ritual (see Tip below).

Also, don't forget to reserve some of the chicken and carrots to make my Healing 'Hug-in-a-bowl' Chicken Soup (see page 95) to make your ingredients stretch.

Serves 4

1 medium-sized organic chicken
½ lemon
5 garlic cloves
Generous pinch each of sea salt and freshly ground black pepper
2 rosemary springs

SPOTLIGHT:
Garlic contains anti-fungal and antibacterial compounds and is full of blood-purifying and immune-strengthening properties. To get the most health benefits from your garlic, chop and grate it about 10 minutes before cooking to increase the allicin (the main active compound) content, so it works extra hard for your body.

1. Preheat the oven to 190°C.
2. Stuff the chicken with the halved lemon and 4 of the garlic cloves.
3. Rub one garlic clove into the skin of the chicken, now rub the skin with sea salt and pepper and place the rosemary sprigs in between the legs. Cover the whole chicken loosely with parchment paper.
4. Place in the oven for 1½–2 hours, until the juices run clear when pierced with a knife.
5. Leave the chicken to rest loosely covered with the parchment paper. Before serving, pour the juices from the roasting tin into a jug to use as your nutrient-dense gravy.

TIP: Shred your garlic-rubbed roast chicken and serve with leafy greens, asparagus, beetroot and buckwheat. Drizzle with a probiotic dressing by mixing 1 tbsp raw honey, 4 tbsp apple cider vinegar and 1 tbsp wholegrain mustard for a spring/summer take on a weekly ritual.

Turmeric Messy Eggs and Avocado

Eggs are a complete protein source and one of the most nutrient-dense foods on the planet – a real nutritional powerhouse. They are a true hormone-balancing superfood, but be sure to eat the yolk, otherwise you will miss out on omega-3, vitamin D, selenium and choline (particularly beneficial for fertility, supporting egg quality for overall thyroid health). Eggs are also a great source of HDL cholesterol, which is the healthy cholesterol your body needs for creating hormones.

I usually make this as a weekend brunch to use up leftover vegetables from the week, or if I get home late and need something quick, healthy, protein-rich and not too heavy to renourish my body systems with. Don't forget to serve with some kimchi or sauerkraut for their gut-feeding probiotic superpowers.

Serves 1

2 or 3 eggs
½ tsp ground turmeric
Pinch each of fennel seeds, Italian seasoning, sea salt and black pepper
Knob of grass-fed butter
2 mushrooms, chopped into small chunks
1 spring onion, finely chopped
¼ pepper (any colour), chopped
1 tbsp chopped sun-dried tomatoes
Small handful of flat-leaf parsley
Big handful of spinach
½ avocado, sliced
1 tbsp sauerkraut or kimchi, to serve

1. Crack the eggs into a jug, add the turmeric, fennel seeds, Italian seasoning, sea salt and pepper and whisk together.
2. Heat the butter in a pan over a medium heat and sauté the mushrooms, spring onion, pepper and sun-dried tomatoes, until soft.
3. Throw in the parsley and spinach, then pour in the eggs, scrambling them in. Keep moving the eggs around the pan, until cooked to your liking. Serve with sliced avocado and sauerkraut or kimchi.

'Friday Night' Chicken Tikka 'Fakeaway'

This gives 'fakeaway' vibes, but will leave you feeling nutritionally balanced instead of depleted and dehydrated the next day. It is also ideal for hosting – it's an easy crowd-pleaser – or to enjoy for a nourishing date night in. It will support your blood-sugar levels as it's high in complete protein to prevent energy dips, and it's big on hormone-healthy nutrients from the organic chicken, including B vitamins necessary for optimal fertility and for nervous-system function (the two go together). The variety of fragrant spices used in the tikka paste adds a dose of micronutrients, too, with powerful female health benefits.

Serve with Greek-yoghurt Flatbreads (see page 200), Protein-rich Fresh Herby Tzatziki (see page 58), a dollop of Greek or natural yoghurt, Quick Pickled Onions (see page 200) and some shredded red cabbage and carrot to help metabolise oestrogen.

Serves 2

500g organic boneless chicken thighs
400g chickpeas, drained and rinsed

For the chicken tikka paste
4 tbsp extra-virgin olive oil
1 tsp ground coriander
1 tsp ground cumin
2 tsp garam masala
1 tsp medium curry powder
1 tsp paprika
Pinch of ground turmeric
7½cm piece of fresh ginger, grated
3 garlic cloves, grated
1 fresh red chilli
1 fresh green chilli
1 tbsp tomato purée
Pinch of sea salt

1. Combine all the ingredients for the tikka paste, coat the chicken thighs with it and leave to marinate for at least 1 hour.
2. Preheat the oven to 200°C.
3. Transfer the chicken thighs to a roasting tin, along with the chickpeas, then roast the thighs for around 45 minutes.
4. Serve with brown rice, quinoa or inside flatbreads with tzatziki.

Favourite Fig and Butter-bean Greek Salad

A dynamic dish that will become a staple and will take the famous summer Greek salad as you know it to the next level, amplified with more plant power so that every system in your body will thank you for it. This is my all-time favourite summer salad recipe. It takes me back to gorgeous memories of being in Greece for family holidays with Sebby. I like to think of it as a summer holiday in a bowl! It really ticks all the boxes: it's slightly sweet, salty, substantial and, to top it off, this one recipe contains a whopping sixteen plant varieties for a happy, well-fed gut microbiome (including key vitamins from the juicy seasonal figs and high-fibre beans to keep your bowels flushing out any excess oestrogen to keep you hormonally balanced and plant protein quinoa to keep you energised and sustained).

This recipe is especially great to enjoy during your ovulatory phase, when fresh, crunchy and antioxidant-rich are what your hormones will thrive most on.

Serves 3–4

250g quinoa
1 avocado
1 green pepper, chopped
1 red pepper, chopped
Big handful of baby spinach
½ cucumber, chopped into small chunks
½ red onion, finely chopped
400g tin butter beans, rinsed and drained
4 figs, sliced, skin on
Big handful of mixed fresh herbs (basil and flat-leaf parsley work well)
2 tbsp black olives
Handful of crushed walnuts
100g feta
2 tbsp hummus

For the dressing

1 tsp Dijon mustard
1 garlic clove, grated
3 tbsp extra-virgin olive oil
3 tbsp white wine vinegar
1 tsp dried oregano
Pinch each of sea salt and freshly ground black pepper

1. Cook the quinoa according to the packet instructions.
2. Combine all the dressing ingredients and set aside.
3. Chop the avocado into small chunks, then transfer, along with the peppers, spinach, cucumber and red onion, to the biggest bowl you have.
4. Add the butter beans, figs, fresh herbs, olives and crushed walnuts.
5. Crumble in the feta, drizzle with the dressing, toss together and serve immediately with some hummus on the side.

TIP: This is best eaten on the same day. If making to use over multiple days, I'd recommend adding the figs and avocado at each sitting.

Orangey Kale and Feta Lentils

If you are struggling with fatigue, heavy periods or low iron stores, this one is for you! I created this nutrient-rich, summery salad for my clients with low iron to help replenish whatever is lost through their menstrual blood. It's high in vitamin C, too, which is proven to increase iron absorption.[87]

Being a cruciferous vegetable, the raw kale supports the breakdown of used hormones, preventing them from recirculating and pushes them out of your body. And I've chosen dill to amplify this recipe, as it's rich in vitamin A, an essential nutrient for a healthy immune system and for optimised reproductive health.[88] The very definition of a cycle-supportive salad!

Serves 2 as a base for your protein source (fillet of salmon, cod, tofu, tempeh or chicken)

1 big handful of kale
1 avocado
400g tin lentils, drained and rinsed
1 Romano pepper
150g feta
4 tbsp pumpkin seeds (part of the follicular phase seed-cycling seeds)
15g dill , chopped
3 tbsp hummus

For the orangey vinaigrette
3 tbsp extra-virgin olive oil
1 tsp Dijon mustard
1 tbsp honey
Pinch of sea salt and freshly ground black pepper
1 tbsp grated orange zest, plus the juice from ¼ orange

1. Put the kale, avocado, lentils, pepper, feta, pumpkin seeds and dill in a big bowl.
2. Combine all the dressing ingredients, then drizzle over the salad, along with the hummus and give it a good mix before serving. Simple!

Love-you-back Greens and Beans Soup

The souper-charged (see what I did there?) hearty soup that will love you right back! This yummy recipe is packed full of immunity-boosting goodness from the greens, ginger and garlic, prebiotics from the miso, hormone-building fats from the coconut milk and fibre from the butter beans. A bowl of love and nourishment. Plus, it keeps in the fridge for three days and freezes well, too.

Serves 3

2 leeks
100g kale
100g broccoli
1 courgette
1 onion (red or white)
1 tbsp extra-virgin olive oil
5cm piece of fresh ginger, grated
4 garlic cloves, grated
½ tsp ground turmeric
½ tsp cayenne pepper
Pinch each of sea salt and freshly ground black pepper
400ml coconut milk
100ml vegetable stock
2 tbsp tamari
1 tbsp miso paste
400g butter beans, drained and rinsed
150g frozen peas
Juice of ½ lemon
Big handful of fresh coriander (approximately 30g)
Mixed seeds (to serve)

1. Chop the leeks, kale, broccoli, courgette and onion.
2. Heat the oil in a big pot over a medium heat, add all the vegetables and cook, until soft (approximately 5 minutes)
3. Add the ginger, garlic, turmeric, cayenne, sea salt and pepper. Pour in the coconut milk and vegetable stock, along with the tamari and miso paste.
4. Add the butter beans, peas, lemon juice and coriander and give everything a good stir, then leave to simmer over a medium/low heat for 10 minutes.
5. Set aside to cool, then transfer to a blender and blitz until smooth.
6. Serve with a sprinkling of mixed seeds on top.

Immune-boosting Antioxidant Salad

The base of this crunchy antioxidant salad is the nutrient-dense showstopper red cabbage, a go-to when I need to give my liver some extra love, whilst also hugely supporting my immunity. This is because red cabbage is rich in anthocyanins, an antioxidant that fights inflammation, free radicals and chronic disease, and it also has more vitamin C than oranges! This recipe also contains pectin – a rich source of prebiotic fibre that feeds the beneficial bacteria in your gut and, as you know by now, a thriving gut is key to better health. The red onion has powerful anti-histamine properties ideal to support allergies.

Serves 2

200g quinoa
¼ broccoli head, chopped into very small chunks
½ small red cabbage head, sliced
½ red onion, finely chopped
200g chickpeas, drained and rinsed
1 Romano pepper, finely chopped
Handful of cherry tomatoes, halved
2 tbsp crumbled feta
2 tbsp crushed almonds
2 tbsp pomegranate seeds
Handful of fresh coriander, chopped

For the Dijon dressing
1 tsp Dijon mustard
1 garlic clove, grated
3 tbsp extra-virgin olive oil
3 tbsp white wine vinegar
Pinch each of sea salt and freshly ground black pepper

1. Cook the quinoa according to the packet instructions.
2. Add the broccoli, red cabbage and red onion to the biggest bowl you have. Pat dry the chickpeas and add them to the bowl, along with the cooked quinoa Romano pepper and tomatoes.
3. Make the dressing by combining the ingredients in a bowl, then drizzle it over the salad and mix well.
4. Crumble the feta and crushed almonds over the top, sprinkle with pomegranate seeds and chopped coriander and dive in!

SPOTLIGHT:
Red cabbage is rich in vitamin K, whch helps maintain bone calcium, therefore a great osteoporosis preventing food.

Beetroot and Parsley Liver-support Juice

This vibrant, liver-loving juice is packed with a powerhouse of essential nutrients to boost your liver's detoxification pathways, which helps with hormone clearance for balanced hormones.

This liver tonic is especially ideal to drink after a night of drinking alcohol, as beetroot and parsley activate your liver's enzymes, which increases bile production and helps your liver detoxify even more effectively to speed up recovery time.

Beetroot also contains pectin, a form of soluble fibre, which helps further flush your system and supports your lymphatic system, too, making it a great skin-support drink.

Serves 2

4 beetroot, chopped
1 cucumber
Juice of 1 lemon with skin on (wash first)
1 apple
5cm piece of fresh ginger, roughly chopped
Big handful of flat-leaf parsley
100ml coconut water (or filtered water if not)

1. Place all the ingredients in a juicer if you have one and serve. If using a blender, simply drain the juice through a nut-milk bag or a sieve to remove the pulp and discard.

TIP: To maintain balanced blood-sugar levels, drink your juice *alongside* a high-protein breakfast and see it as an antioxidant bonus on the side (drinking juice as a meal replacement will negatively impact your blood-sugar levels because the fibre has been removed).

Rainbow Summer Tofu and Quinoa Spring Rolls

Soul food that makes you smile from the inside out! Ideal for those warm, sunny, follicular or ovulatory days when you need all the antioxidant-rich freshness you can get from nature's produce but fancy a change from salad. It's packed with vitamin-rich, colourful, crunchy raw veggies and your gut and liver will thank you for helping to metabolise excess hormones to promote balance. Plus, crispy complete-protein tofu and quinoa support your blood-sugar levels and keep you satisfied. Finally, my addictive satay dipping sauce completely brings these rainbow rolls to life. They're more filling than they look and are ten out of ten on taste! Enjoy crunching and dipping your way through a batch of these.

Serves 2

150g quinoa
200g organic extra-firm tofu
Drizzle of extra-virgin olive oil
2 tsp tamari
½ red pepper
½ yellow pepper
2 spring onions
¼ red cabbage
1 carrot
⅓ cucumber
1 avocado
6 spring-roll sheets
Handful of mint or coriander (or a mix of the two)

For the satay dipping sauce

2 tbsp smooth peanut butter
2 tbsp tamari
1 tbsp toasted sesame seed oil
1 garlic clove, grated
4 tbsp water (to loosen as needed)

1. Cook the quinoa according to packet instructions.
2. For the tofu, start by pressing it between some kitchen paper to squeeze out any excess moisture, then chop into small rectangle shapes.
3. Heat the extra-virgin olive oil in a pan on medium heat, add the tofu and toss in the tamari, flipping the tofu on all sides to ensure even browning (this should take approximately 5 minutes). When the tofu is golden and crispy, remove from the pan and set aside to drain on kitchen paper.
4. Finely chop the peppers, spring onion, red cabbage, carrot, cucumber and avocado into super-thin strips (the thinner, the better).
5. Place each spring-roll sheet in a shallow bowl of warm water and submerge for about 15 seconds, until soft and pliable. Shake off any excess water and place the sheets on a tea towel, then very gently pat away any remaining water.
6. Divide the tofu, cooked quinoa and thinly sliced vegetables between the spring-roll sheets. Fold over the top and bottom, then roll to finish.
7. Mix the satay sauce ingredients together, adding enough water, gradually, to achieve a dipping consistency. And that's it – super simple and extremely tasty!

Creamy Cauliflower, Leek and Cashew Soup

Leeks are your liver's friends, as they're super high in sulphur compounds, which help to trap and flush away toxins, accelerating your detox and helping to keep you balanced from the inside out. Plus, if you suffer with water retention, this recipe is the ideal remedy, as leeks are a diuretic helping to naturally rid the body of excess fluid.

Stirring in some precooked quinoa or buckwheat to serve makes this soup even more sustaining.

Serves 4

1 tbsp extra-virgin olive oil
2 leeks, chopped
4 garlic cloves, chopped
1 medium-sized cauliflower head, chopped
100g cashew nuts
Pinch of smoked paprika
Pinch each of sea salt and freshly ground black pepper
1 litre vegetable stock
30g flat-leaf parsley
Mixed seeds, to serve

1. Heat the oil in a pan and sauté the leeks and garlic for a few minutes, before adding the cauliflower and cashews.
2. Add the paprika, salt and pepper, then pour in the stock and simmer for 20 minutes.
3. Transfer the soup to a blender along with the parsley and blitz, until smooth.
4. Sprinkle with mixed seeds.

Liver-detox Shredded Asian Satay Salad

This is not just any salad. It's a 10/10 statement salad full of diversity and brought to life with a zingy satay sauce that you will keep coming back to time and time again, it's that good!

Loaded with liver-loving vegetables, such as antioxidant quercetin, high in sulphur-containing amino acids, which help to promote liver detoxification, this raw salad makes a delicious main meal, accompanied with salmon, prawns or tofu or it's great on its own for a transportable lunch.

Serves 4

½ red cabbage head
2 medium carrots
1 Romano pepper
1 red onion
2 servings of soba noodles or 2 gluten-free wholegrain rice noodle 'nests'
3 tbsp kimchi
100g organic edamame beans
Handful of fresh coriander (approximately 30g)
2 tbsp black sesame seeds (skip this if seed cycling)
1 tbsp white sesame seeds (skip this if seed cycling)
20g pistachios, chopped

For the zingy satay sauce
2 tbsp peanut butter
2 garlic cloves, grated
Juice of 1 lime
2 tbsp cold-pressed toasted sesame oil
1 tsp Chinese rice vinegar
Pinch of dried chilli flakes
30ml water (to loosen as needed)

1. Finely shred the cabbage, carrots, Romano pepper and onion using a mandolin, grater or food processor and transfer to the biggest bowl you have.
2. Combine all the ingredients for the satay sauce, gradually adding the water and stirring well to achieve a runny consistency.
3. Cook the soba or wholegrain rice noodles according to the packet instructions (or you could sub this for quinoa).
4. Add the noodles to the bowl, along with the kimchi, edamame beans and coriander.
5. Drizzle the satay sauce over the shredded veggies and noodles, sprinkle with sesame seeds (if using) and chopped pistachios and enjoy!

Liver-loving Ginger and Lemongrass Sea Bass

Lemongrass is an underestimated liver-lover. Studies show that lemongrass can reduce oxidative stress in the liver as it contains an abundance of inflammation-fighting compounds and is also a good source of quercetin (also found in red onion), a flavonoid known for its powerful antioxidant and anti-inflammatory benefits.[89]

Much like lemongrass, grated ginger further amplifies this nourishing meal, enhancing your digestion and strengthening immunity and the high protein content from this recipe (from the sea bass and edamame) means it will help your liver produce glutathione, an extremely important antioxidant, vital for liver function.

I like to serve this superfood sea bass with black rice. Research shows that black rice contains over twenty-three types of antioxidants and has the highest antioxidant activity of all rice varieties.[90]

Serves 2

2 servings of black or brown rice (approximately 250g)
2 sea-bass fillets
150g organic edamame beans
Drizzle of extra-virgin olive oil
2 pak choy
1 garlic clove, grated
1 tbsp sesame seeds or crushed cashews

For the marinade

3 tbsp tamari
2 tbsp raw honey
1 thumb-sized piece grated ginger
⅓ lemongrass stalk grated

1. Preheat the oven to 190°C.
2. Cook the rice according to the packet instructions.
3. While the rice is cooking, combine all the marinade ingredients.
4. Place the sea-bass fillets on a non-stick baking tray or line a tray with baking paper and coat with the marinade (reserving a little for the pak choy). Roast in the oven for around 15–18 minutes (check halfway through and quickly baste with the marinade).
5. Around 5 minutes before serving, steam the edamame beans, so they are still bright green but starting to soften (around 4 minutes).
6. At the same time, heat the oil in a pan over a medium–high heat and add the pak choy, along with the reserved marinade and grated garlic. Toss and cook until the pak choy is nicely wilted.
7. Serve the fish with the rice, pak choy and edamame beans and sprinkle with sesame seeds or crushed cashews to finish.

TIP: Save some lemongrass for a delicious anti-inflammatory tea or to make a change to your regular lemon and ginger in the morning.

Ultimate Oestrogen Detox Reset Salad

During the latter part of your follicular phase and your ovulatory phase, your body needs support clearing oestrogen to stop too much circulating your system, to prevent PMS symptoms and promote cycle homeostasis. This is where two of my favourite hormone-clearing superheroes come in – the humble carrot and broccoli, the base of this detox salad. Carrots contain a unique fibre called lipopolysaccharides (LPS), which binds to endotoxins and connects to hormone imbalances linked to an excess in oestrogen. The high fibre content of raw carrot also binds to excess oestrogen and safely eliminates it from your body via your bowels to promote balance. The broccoli is rich in a sulphur-rich plant compound called sulforaphane. We activate this compound when we chop or chew brassicas, and it is at its highest when eaten raw. It is fantastic at promoting the clearance of oestrogen, helping you feel detoxed in the most balanced way.[91] No fad detox juices here!

Serves 2

200g cooked quinoa
200g raw broccoli, finely chopped
2 carrots, chopped into chunks
1 apple, chopped into chunks
¼ red onion, chopped into chunks
Handful of cherry tomatoes, chopped into chunks
1 avocado, chopped into chunks
½ cucumber, chopped into chunks
4 small radish, finely chopped
200g chickpeas, drained and rinsed
Small handful of crushed walnuts
Handful of flat-leaf parsley (approximately 30g), finely chopped

For the detox dressing
1 garlic clove, grated
2 tsp raw local honey
3 tbsp extra-virgin olive oil or avocado oil
1 tbsp apple cider vinegar
1 tsp za'atar
Pinch each of sea salt and freshly ground black pepper

1. Add the quinoa, broccoli, carrots, apple, onion, tomatoes, avocado, cucumber and radish to a big bowl. Pat dry the chickpeas with a tea towel and add to the bowl of chopped vegetables.
2. Combine all the ingredients for the Detox Dressing.
3. Drizzle the salad ingredients with the dressing, toss together, top with the crushed walnuts and the parsley and dig in!

The Balance Pot

I've been making these little pots for years. It's a convenient way to nourish your body with the perfect balance of complete protein, healthy fats and complex carbs, plus liver-loving bitter greens, prebiotic-rich spring onion, and an array of nutrient-dense seeds. Enjoy at home or as an on-the-go breakfast to set you up for the day ahead. Your hormones will quite literally be thanking you for feeding them with this powerful, balanced little combo. The pots can be served warm in luteal and menstrual phases, or cold in follicular and ovulatory phases.

Serves 1

3 tbsp buckwheat
2 organic eggs
½ avocado
½ spring onion, finely chopped
Small handful of bitter greens (spinach leaves, rocket and parsley)
1 tbsp mixed seeds or your seed- cycling seeds
Pinch of fennel seeds
3 walnuts, crushed
Drizzle of extra-virgin olive oil
Optional extras: pinch of sea salt, pepper, small drizzle of balsamic vinegar, chilli flakes

1. Cook the buckwheat according to the packet instructions (I do this the night before to save time).
2. Boil the eggs for 9–10 minutes. Once cooked, remove from the pan and transfer to a bowl filled with ice and water for a few minutes – this will make it much easier to peel.
3. In the meantime, add the avocado half and spring onions to a bowl with a little handful of the greens.
4. Cut the eggs in half and place in the bowl with your chosen seeds, fennel seeds and walnuts. Drizzle over some extra-virgin olive oil, and finish with some seasoning, plus a little balsamic if you fancy a bit of sweetness and chilli flakes for a touch of heat!

TIP: If making your pot the night before to save time in the morning as a grab-and-go breakfast, remember to leave the stone of the avo in to prevent it going brown.

6. The Skin

'What manifests on the inside, shows up on the outside.'

To kick this chapter off, I'm reminding you that your skin is your body's largest organ, protecting your precious internal organs and systems from the outside world. Healthy skin is, therefore, so much more than just a vanity project (although that, by the way, is absolutely fine, too – who doesn't want gorgeous, glowing skin?). What's more, healthy skin is a visual indicator of good overall health and balanced hormones.

Balance Your Hormones, Balance Your Skin

Hormonal acne might make you think of being a spotty, sassy teenager (or maybe that was just me!), but the reality is that it can impact you at any point during your adult life from menstruation to menopause, again due to hormonal fluctuations (usually reduced levels of oestrogen).[92]

Hormonal acne refers to breakouts that occur due to hormone fluctuations – as your hormone levels change, so does the amount of oil your skin produces. This is particularly connected to the rise in androgens, such as testosterone, but is also associated with general hormone imbalance. Remember the symphony orchestra analogy earlier (see page 28)? Well, when one hormone is out of whack it can easily cause hormonal havoc, and this can quite literally show up on your face.

When it comes to a rise in androgens, the biggest cause of hormonal acne, it's the rise in testosterone levels that can trigger the excessive secretion of sebum from glands in the skin. It may also change the activity of certain skin cells, leading to an infection of the hair follicles by a bacterium known as *Cutibacterium acnes*, which can contribute to hormonal acne. Your body's immune system may react to the bacteria and its metabolites and produce inflammation, causing redness alongside acne lesions.

Fluctuations in the levels of oestrogen and progesterone during your cycle can also be contributors to unhappy skin. Research on this is inconclusive, but the hypothesis is that increased progesterone raises your body temperature, leading to more sweating and clogged pores. The build-up of sebum beneath the skin's surface, along with dirt, dead skin cells and bacteria can cause breakouts of acne before and during your period. I remember my hormonal acne like it was yesterday: it would angrily appear along my jawline and on my chin, paired with tender breasts and anxiety – my body's way of telling me something was out of balance internally and needed nourishment. If this resonates with you, I hope you will start to take your skin's needs seriously and recognise how it's communicating with you. You don't have to put up with 'bad' skin, and I'm here to remind you that it doesn't have to be your normal.

Eat Your Skincare: Inside-out Beauty

Yes, applying the right skincare products with the right ingredients, tailored to your skin type and concerns is important – and I'm the biggest skincare sucker out there. But you cannot cleanse, serum and facial your way to your best skin in isolation, without making fundamental changes to what you put *inside* your body. And this starts with a focus on the skin nutrition you put on your plate at

each meal and being savvy about the skin-depleting foods you should minimise.
As clichéd as I know it sounds – and I'm aware you've heard this a million times before – good skin really does start from the inside, which is why in this chapter, we're going to touch on the gut–skin axis, detoxification pathways and how hormone imbalances are at the root of your skin issues. Because, as you know by now, getting to the root of your symptoms and using food is the *only* way to create transformative and lasting change.

I'm here to help you eat your way to better health, which means healthy, glowing and, most importantly, happy skin. So it's time to start upping your antioxidants (in the form of berries and brightly coloured vegetables, including sweet potatoes, squash and beetroot), embracing fatty acids (from oily fish, avocados, nuts and olive oil) and eating a rainbow of high-fibre foods at each and every meal to keep your system topped-up with the vitamins and minerals it needs to be clear and glowing and to flush out the toxins that contribute to your skin concerns.

Skin Symptoms: Signals of Gut and Hormonal 'Malfunctions'

You've probably noticed that when you experience other body system and hormonal 'malfunctions', as I refer to them, such as constipation and PMS, that your skin is also negatively impacted. This may show up as breakouts, or for you it could mean skin that's dull and lacklustre in appearance, 'bumpy' to touch, reactive or sensitive, or perhaps you've noticed dark circles or accelerated ageing (formation of new lines/wrinkles) or even cellulite that wasn't there before. These are visible signs and signals that something is out of balance internally, so don't ignore them. And don't forget, when we talk about skin, it's not just the skin on your face that's impacted; the skin on your whole body is communicating with you, dictated to by your hormones and nutrition.

YOUR SKIN IS A MIRROR TO YOUR GUT (#GUTSKINAXIS)

You might be starting to wonder whether this should, in fact, be the gut rather than the skin chapter. But when it comes to our skin, starting with the gut is both wise and essential.

I like to think of the skin as a mirror to the gut because of its clever ability to reflect what is going on internally. The intimate relationship between these organs is now widely known as the 'gut–skin axis', a beauty buzzword that refers to the communication between the gut and the skin and how the health of one impacts the other, showing that to successfully support the skin, we must start by nourishing the gut.

ARE YOU POOING EVERY DAY?

When working in practice with clients who come to see me for nutritional support for their acne, rosacea or 'hormonal' breakouts, the very first area of their body I ask them about is their gut (more specifically, their digestive system and bowel habits), as the two are intrinsically linked. I ask how many times they poo each day and if they are fully emptying their bowels. This is because, if you are either not pooing regularly (and when I say regularly, I mean at least daily) or your bowels are on the sluggish side, then your skin will start to look as congested your insides are.

MALABSORPTION AND MALDIGESTION: THE ENEMY OF GOOD SKIN

If your gut is not playing ball and is not as healthy or functioning as well as it should be, then you must consider malabsorption (when the digestive system prevents the body from absorbing nutrients from food) and maldigestion (when the digestive system is not breaking food down effectively enough to extract the nutrients you need), as these can reduce the amount of nutrients available to your body.

This lack of nutrients can cause unhappy skin. For example, if you're not absorbing enough skin-essential vitamins (such as vitamins C and E, both of which are vital for collagen formation and skin repair), zinc (for healing breakouts, including acne), fatty acids (which regulate the skin's oil production, keeping it hydrated and keeping inflammation at bay) and protein (remembering that protein is literally the building block of your skin, hair and nails), you can understand how this could lead to problems with how your skin looks, feels and behaves.

Inadequate nutrient absorption can also negatively impact the production of hormones. For example, if you're not getting enough selenium, iodine, vitamin D and B vitamins, all of which are required for good thyroid health, this could create hormonal imbalances. And this, too, can affect the health of your skin.

If you are concerned you are not absorbing nutrients properly and your gut needs support, my suggestion is that you work with a naturopathic nutritionist who can tailor a protocol to your individual needs. But in the meantime, if you follow the recipes here and in the gut chapter (see page 73–109) and begin nourishing your body to better health, your skin will reap the benefits

More Fibre + More Varieties = Healthier Skin

Your first port of call when it comes to eating to support your skin is to increase the amount of dietary fibre you eat from a rainbow of phytonutrient and antioxidant-rich wholefood plants. These plants will reduce the inflammation and oxidative stress that drive/worsen 'angry' skin. Aim for 30g per day (the UK average is approximately 14g per day), as this will optimise the health of your skin microbiome and, in turn, regulate your bowel movements – the two go hand in hand. To make this easier, and because I follow an intuitive-eating approach and therefore don't like to weigh foods or focus on numbers (see page 16), I try to eat half a plate or bowl of 'rainbow' veggies at every meal. I also make a conscious effort to switch up the varieties of plants in my weekly meals, as this exposes gut microbes to different sources of fibre (studies have shown that a greater dietary fibre intake is associated with increased gut microbe diversity), which produce short-chain fatty acids to promote a healthy colon.

Good sources of dietary fibre include oats, chia seeds, flaxseed, all beans and pulses, grains, root vegetables, avocados, apples, bananas, nuts and seeds. Including pre- and probiotics in your diet also increases the number and diversity of the bacteria in the gut, which can help to reduce the growth of pathogenic microbes living there, aid digestion, boost the helpful chemicals that gut bacteria produce (more on this to follow) and prime your immune system.[93] All of which contributes to healthy skin.

Getting to the Root of Your Skin Issues

If your skin has more 'bad' days than 'good/normal' ones and other malfunctions are also causing you problems, then it's time to renourish your skin back to full, glowing health, using food as your medicine. This will not just improve your skin and how it detoxes on the inside and appears on the outside but will help with associated symptoms by getting to the root of the issue, as we are never eating to treat just one thing.

Good Detoxification Means Good Skin

Unhealthy skin is usually down to a combination of imbalanced hormones and your detoxification pathways struggling and asking to be optimised, often driven by poor diet and too much toxin exposure (see EDCs on page 113). This can cause issues with your digestive health and can signal that your liver, kidneys, lungs and lymphatic systems could do with some extra nourishment and TLC.

Goodbye, Restrictive 'Detoxes'; Hello, Daily Nourishment

I often see people doing restrictive cleanses and 'detoxes', but inevitably going back to the foods that drove their skin imbalances in the first place. Trust me, I've been there many times: I've drunk the overpriced juices and starved myself of nourishment and, ultimately, learned that these well-marketed cleanses were never going to give me the skin results I wanted. Instead, they just drove my hormonal imbalances deeper and made me crave unhelpful foods more than ever.

Don't get me wrong – there are multiple factors at play in good skin. But by using real food, grown and gifted by nature, to support your body's natural ability to detox each day of the year, you will find a sustainable (and tasty) way to give your skin the continuous support it craves – and it will repay you in 'glow', which you will see in the mirror and people will compliment you on. I often think when someone says, 'You look well', they are noticing that someone looks healthy and glowing.

EXCESS HORMONES NEED SOMEWHERE TO GO

Good detoxification pathways mean good elimination pathways, which result in good skin. When these are not able to carry out their jobs for you in the way they've been designed to do, this will show up on your face (and/or the skin on your body).

Good elimination essentially supports hormone metabolism (the breakdown of hormones), so that used-up hormones and toxins are not recirculated in your blood, which would be a drain on your mitochondrial cells (the body's battery pack) and could drive inflammation, which, yep, you guessed it, can manifest in your skin. (See how interconnected your body systems and hormones are?) But when your detoxification systems are all running as they should, the unhelpful stuff is being flushed efficiently from your body.

Remember, toxins can come from the beauty and household products you use, which, of course, you can change (for more on endocrine disruptors, see page 113), although you cannot always control your exposure to environmental toxins. But what you eat is within your control and should always be your starting point when it comes to improving your body systems and organ function, so they can work their magic for you and your skin.

Skin-feeding Foods

Specific foods have the power to upregulate your metabolic pathways to assist with toxin elimination, which will help you on your journey, and these foods are woven into the recipes I've created in this chapter to nourish your skin.[94]

Cruciferous vegetables: broccoli, cabbage, kale, pak choy, collard greens, watercress
Herbs: parsley, thyme, dandelion, oregano, coriander, rosemary
Spices: ginger, turmeric
Fruits: berries, avocados, lemons, limes, oranges
Oily fish: wild salmon, mackerel
Root vegetables: sweet potatoes, butternut squash, beetroot, garlic
Nuts and seeds: walnuts, almonds, Brazil nuts, pumpkin seeds, flax, sunflower (your seed-cycling seeds – see page 40–43).

You will notice that many of these ingredients appear in in the recipes that follow. I've included them to provide your skin with nutrients from a range of phytochemical-rich plants and nutrient-dense foods brimming with the antioxidants, antimicrobial and anti-fungal properties, healthy fats and fibre required to nourish your skin to better health.

The Nourish Method Skin Reset Ritual

At the end of each day, perform my skin reset ritual as part of your skincare routine. It only takes 5 minutes but it's one of the most relaxing yet powerful ways to physically support your skin, alongside using food as your skin medicine.

Facial massage enhances lymphatic flow and detox (promoting toxin elimination for clearer, healthy skin), improves blood flow for visibly glowing skin, releases muscle tension to make the appearance of fine lines less prominent and stimulates collagen production which keeps skin looking firm and plump. As a bonus, it also activates your parasympathetic nervous system making it a great habit-stack routine to help wind-down after a busy day and gift the nervous system some downtime.

Cleanse your complexion, gently pat dry, moisturise and then apply a facial oil to carry out the skin reset ritual (applying them in this order locks in the moisture):

Forehead: Place your knuckles between your eyebrows and move them up towards the hairline. Repeat this five times.

Cheeks: Rest your knuckles on your cheeks near the nose bridge and swipe them across your cheeks towards the ears. Repeat five times.

Mouth: Make a 'V' shape with your index and middle finger. Place the fingers in a way that the index finger is above the upper lip and middle finger is below the lower lip. Press gently and pull towards the ears.

Chin: Place your palms underneath your cheekbones and massage upwards.

Eyes: Use soft and rolling movements with your ring finger. Repeat five times.

Or you could use a gua sha tool to massage your face. Follow the above instructions as a guide but replace the knuckle and finger instructions with the gua sha instead.

Skin Recipes

Peanut and Ginger Noodle Stir-fry

Making your own sauces and marinades from just a handful of store-cupboard staples (plus a few fresh anti-inflammatory ones like ginger and garlic) is a great way to reduce unnecessary chemicals in your diet. Not to mention it's cost-effective, too, as you can use the ingredients you buy for multiple recipes in this book. And you can also buy pre-prepared mixed stir-fry veggies to save time.

I've been making this peanut stir-fry sauce for years. It will totally elevate your speedy weeknight stir-fries, transforming a bunch of plain veggies into a taste sensation.

Serves 2

2 gluten-free wholegrain noodle nests or 2 servings soba noodles
100g fresh or frozen organic edamame beans
1 tbsp extra-virgin olive oil
2 mixed colour peppers, chopped
1 large red onion, chopped
150g pak choy, chopped
150g tenderstem broccoli, chopped
1 tbsp black sesame seeds
1 tbsp white sesame seeds
1 spring onion, finely chopped
Handful of chopped cashews
2 lime wedges
Big handful of fresh coriander (approximately 30g), plus extra to serve

For the peanut stir-fry sauce

5cm piece of fresh ginger, grated
4 tbsp tamari
3 garlic cloves, grated
2 tbsp smooth peanut butter
1 tsp maple syrup
1 tbsp toasted sesame oil
1 tsp Chinese rice vinegar
Juice of 1 small lime
30ml water to loosen (add slowly to achieve desired consistency)
Generous pinch of dried chilli flakes

1. Prepare the peanut stir-fry sauce by combining the ingredients together in a bowl.
2. Cook the noodles according to the packet instructions.
3. If using frozen edamame beans, cover with boiling water to thaw. Set aside.
4. Drizzle the extra-virgin olive oil into a pan over a medium heat. Add the chopped stir-fry vegetables to the pan and sauté. Be sure to keep moving them around pan and don't allow them to burn.
5. When the vegetables are soft, drain the edamame beans and noodles and throw them into the pan.
6. Pour the peanut sauce into the pan over the vegetables, sprinkle over the sesame seeds and combine.
7. Divide between two bowls and top with the spring onion, chopped cashews, a squeeze of lime and coriander.

TIP: Add your choice of protein (I like prawns or tempeh with this sauce) or double up on the edamame. If using prawns, fold them in just before step 7 and keep moving them around the pan, until cooked (just a few minutes if using precooked or longer if using uncooked – about 4 minutes). If using tempeh, then cut it into 2cm cubes, add to the pan at the same stage and roll into the sauce to season it. Stir-fry for around 4 minutes, until cooked.

'Sunshine-in-a bowl' Salmon Salsa

'Sunshine' in a bowl is the only way to describe this recipe.

It reminds me of Amy, a friend who's more like a sister to me. Not only is she human sunshine, she also happens to be the first person I made this recipe for and we've been enjoying it as our go-to girls'-night dinner ever since. The rainbow of juicy, fresh ingredients makes it rich in antioxidants and phytonutrients, including oestrogen-metabolising fibre and omega-3 fatty acids to promote hormone balance, boost your mitochondria for increased energy and to leave you feeling satisfied. It's perfect spring/summer nourishment for the follicular and ovulatory phases of your cycle when you crave light, fresh crunchiness.

Serves 2

2 wild salmon fillets
¼ red cabbage head, finely sliced or shredded
Handful of vine tomatoes, sliced
1 Romano pepper, sliced
¼ red onion (or 2 spring onions), chopped
1 avocado, chopped into chunks
1 mango (or a big handful of frozen mango)
Juice of 1 lime
Pinch of sea salt and freshly ground black pepper
2 tbsp extra-virgin olive oil
Handful of fresh coriander, roughly chopped
400g tin chickpeas, drained and rinsed
Handful of crushed walnuts
½ fresh red or green chilli, finely chopped or sprinkle of dried chilli flakes (optional)

For the salmon seasoning

2 tbsp extra-virgin olive oil
1 garlic clove, grated
1 tsp Cajun spice
Small squeeze of lime
Tiny pinch each of sea salt and freshly ground black pepper

1. Preheat the oven to 190°C.
2. Combine all the ingredients for the salmon seasoning. Coat the salmon fillets with the seasoning and pop it in the oven, uncovered, for approximately 12–16 minutes, or until cooked to your liking.
3. In the meantime, in a big bowl add the finely sliced or shredded red cabbage, sliced tomatoes, pepper and red onion (or spring onions). Add the avocado and mango, along with the lime juice, sea salt, pepper, extra-virgin olive oil and coriander.
4. Pat dry the chickpeas, then add to the bowl and mix. Have a quick taste to see if more seasoning is needed.
5. When the salmon is ready, place it on the salad, top with crushed walnuts and green chilli and serve your sunshine in a bowl!

Harissa Cod, Beets and Bean Skin-support Traybake

Beetroot has been ranked as one of the most potent antioxidant vegetables out there. Its impressive phytochemical profile is significant for your skin health because they help your body combat the age-accelerating, inflammatory effects of a process called oxidation (which happens mainly due to sun exposure and pollution) thanks to high amounts of anthocyanins, which are also responsible for their beautiful, vibrant colour.

I love this super-simple traybake recipe because not only is it delicious and I'm obsessed with anything harissa, but it's also quick to throw together, which is always a win in our household! From a nutritional perspective, it provides complete protein for its essential-amino-acid content (healing and repair), fibre for elimination (to support and prevent breakouts) and, as mentioned, phytonutrients to amplify skin health.

The marinade also works well with chicken, salmon or roasted cauliflower.

Serves 2

2 cod fillets
400g tin butter beans (borlotti or chickpeas also work well), drained and rinsed
2 large beetroots, cut into thin circles
200g tenderstem broccoli, chopped
1 red onion, chopped
1 Romano pepper, chopped
4 garlic cloves, chopped
Handful of cavolo nero, chopped

For the harissa marinade
1½ tbsp harissa paste
1 garlic clove, grated
4 tbsp extra-virgin olive oil
1 tsp grated lemon zest
Juice ½ lemon
Small handful of flat-leaf parsley, finely chopped
2 tbsp raw honey
Pinch each of sea salt and freshly ground black pepper

1. Preheat the oven to 200°C.
2. Combine all the ingredients for the harissa marinade and coat the cod fillets with it, reserving 1 tablespoon to mix into the veggies and beans.
3. Pat dry the beans and add them, with the other ingredients (apart from the cavolo nero), to a roasting tin. Mix in the reserved harissa marinade and place the cod fillets in the middle. Roast for 30 minutes.
4. About 10 minutes before the end of the cooking time, add the cavolo nero.

Avocado, Herb and Walnut Skin-glow Sourdough

Eczema, dry or dull skin in need of some glow? This delicious brunch idea delivers major skin-feeding benefits. A prime example of medicine food, avocados are an internal skin moisturiser, rich in fatty acids and vitamin E to reduce inflammation, nourish and promote glow from the inside out.

Spread on sourdough for the probiotics and fermented fibre to support your gut (the key to happy skin) and top with walnuts (another great source of fatty acids), anti-inflammatory nutrient-loaded fresh herbs and a touch of medicinal raw honey for some serious skin food. And why not add a sliced hardboiled egg on top for a complete protein source?

Serves 2

2 slices sourdough, toasted
1 avocado
Small handful each of flat-leaf parsley and basil, roughly chopped
Pinch each of sea salt and freshly ground black pepper
Drizzle of extra-virgin olive oil
Pinch of dried chilli flakes (optional)
Small handful of walnuts, roughly chopped
Drizzle of raw honey

1. Slice and toast the sourdough
2. Mash the avocado in a bowl with the herbs and season with sea salt, pepper, extra-virgin olive oil and chilli flakes (if using).
3. Top with the roughly chopped walnuts and a drizzle of honey.

TIP: Serve with a tablespon of sauerkraut or kimchi for a dose of gut-nourishing probiotics

Quinoa-crusted Aubergine Parmigiana

This has all the unbeatable Italian lasagne vibes, only without the heaviness of the meat and pasta and with quinoa instead – rich in B vitamins and magnesium, which play a pivotal role in regulating hormonal balance. The aubergine base is loaded with antioxidant flavonoids, which help to protect skin cells from damage and transport nutrients such as iron around your body, while the fibre-rich chickpeas work to eliminate toxins from your system.

All in all, this is the ultimate crowd-pleaser recipe that you can make in advance and just pop in the oven when you want to eat. Ideal for hosting, it's a great dish to feed family, friends and little ones, too!

Serves 4

250g quinoa
2 large aubergines, sliced into rounds
1 large courgette, sliced into rounds
1 tbsp extra-virgin olive oil
4 garlic cloves, grated
1½ tbsp balsamic vinegar
2 tbsp red pesto
Big handful of basil, roughly chopped
680g jar passata
2 x 400g tins chickpeas, drained and rinsed
½ tsp fennel seeds
Pinch of sea salt and big twist of black pepper
2 mozzarella balls, roughly torn
Sprinkle of dried chilli flakes (optional)
Parmesan, grated (optional)

1. Preheat the oven to 180°C.
2. Cook the quinoa according to the packet instructions.
3. Place the aubergine and courgette slices on a baking tray, drizzle with the extra-virgin olive oil and heat both sides under the grill until lightly browned. Set aside.
4. Add the garlic to a pan over a medium heat, along with the balsamic vinegar, red pesto and basil.
5. Pour in the passata, rinsed chickpeas, fennel seeds and salt and pepper. Let it simmer for 10 minutes, then remove from the heat.
6. In a large ovenproof dish, form a layer of the aubergine and courgette slices. Cover with a layer of the tomato sauce, then a layer of torn mozzarella. Keep layering until you have used all the vegetables and sauce and most of the mozzarella.
7. Finish with the quinoa, the remaining mozzarella, chilli flakes and Parmesan, if using.
8. Bake in the oven for 30 minutes until golden and crispy on top and enjoy with a big, leafy avocado salad for extra nutrients and healthy fats.

TIP: If you're dairy free, why not swap mozzarella for nutritional yeast between the layers and sprinkled on top?

Love Your Skin Lentil Salad

Everything about this simple salad has skin-glow properties, from the beetroot and parsley, which will support your liver detox, to the carrot, being high in vitamin A, which is important for skin healing and a healthy immune system. I've also chosen a base of prebiotic-rich lentils to feed your good gut bacteria and the kale, for a thriving microbiome because #gutskin connection is vital! Eat this on its own or with your choice of protein.

Serves 2

200g quinoa
400g tin green lentils, drained and rinsed
4 small beetroot balls
2 medium-sized carrots, sliced
¼ red onion, finely chopped
Big handful of kale, roughly chopped
15g flat-leaf parsley, finely chopped
15g mint, finely chopped
Crumbled feta and crushed walnuts/mixed seeds, to finish (optional)

For the glow dressing

3 tbsp apple cider vinegar
Juice of ½ lime
2 tbsp extra-virgin olive oil
Small handful of mixed chopped mint and flat-leaf parsley
2 tbsp maple syrup
1 tsp Dijon mustard (to taste)
Generous pinch each of sea salt and freshly ground black pepper

1. Combine all the ingredients for the Glow Dressing.
2. Cook the quinoa according to the packet instructions and pat dry the green lentils. Add the quinoa to a big bowl, along with the lentils, beetroot, carrots and red onion and mix together.
3. Add the kale, parsley and mint on top, drizzle with the dressing and toss together. Top with the feta and walnuts or mixed seeds (if using) and serve.

Beetroot Beauty–food Bowl

Another prime example of an abundance of beauty food in one tasty bowl! This one contains beans to support elimination pathways, prebiotic pectin from the apples to feed your gut bugs and walnuts for their high plant-based omega-3 content. But the star ingredient has to be the beautifying beetroots – these earthy, antioxidant-rich root vegetables containing collagen-building vitamin C, iron, beta-carotene and lycopene, which promotes skin elasticity and helps protect skin from sun damage.

If beets are not your thing, I urge you to give them a second chance because slowly roasted and caramelised with honey and a pinch of rosemary absolutely elevates them. And knowing their incredible skin benefits should help, too!

Serves 2

4 medium-sized beetroot
Drizzle of extra-virgin olive oil, plus extra for dressing (with a drizzle of balsamic vinegar)
1 tbsp raw honey
1 tsp rosemary
Pinch each of sea salt and freshly ground black pepper
400g tin green lentils, drained and rinsed
Big handful of mixed spinach and rocket
1 apple, cored and finely chopped (unpeeled for extra fibre and nutrients)
½ red onion, finely chopped
100g goat's cheese
2 small handfuls of crushed walnuts

1. Preheat the oven to 190°C.
2. Wash the beetroot, pat them dry and cut into small chunks. Place in a roasting tin and drizzle with extra-virgin olive oil, add the honey and season with rosemary, sea salt and pepper. Roast in the oven for 30 minutes or until perfectly cooked, turning the beetroots halfway through the cooking time.
3. In the meantime, pat dry the lentils and add them to a big bowl with the spinach and rocket, apple, red onion, goats' cheese and crushed walnuts.
4. When the beetroots are ready, add them to the bowl, drizzle with a simple balsamic and olive-oil dressing and serve.

Glowing Green Miso Noodle

The best way to describe this beauty-boosting soup is glow in a bowl. Shiitake mushrooms, garlic and ginger are three key skin-feeding foods with remarkable medicinal powers when it comes to inhibiting inflammation, improving circulation and promoting healing with respect to optimising your skin health.

Shiitake mushrooms, contain several antibacterial, antiviral and antifungal compounds, especially beneficial for treating acne and breakout skin. They also have high amounts of copper and are recognised for their immune-modulating properties due to the polysaccharides they contain. And a healthy immune system is key to healthy skin that glows from the inside out.

Serves 4

200g shiitake mushrooms
1 tbsp quality sesame oil
1 broccoli head, separated into florets and chopped
1 courgette, chopped
2 green chillies, chopped
7½cm piece of fresh ginger, grated
4 garlic cloves, grated
250g pak choy, ends cut off
Big handful of coriander, roughly chopped
3 tbsp tamari (or to taste)
1 litre vegetable stock
2 tbsp white miso paste
1 tsp mild chilli powder, plus pinch of dried chilli flakes (optional)
200g organic edamame beans
3 wholegrain noodle nests
3 spring onions, sliced
Juice of 1 lime
2 tbsp sesame seeds, to garnish

1. Chop the stalks off the mushrooms and then cut them into slices.
2. Heat the oil in a big pot over a medium heat, then add the broccoli, courgette, chillies, ginger and garlic. Stir-fry for a few minutes until slightly soft.
3. Now add the mushrooms, pak choy, half the coriander and the tamari.
4. Pour in the stock, add the miso paste, chilli powder and edamame beans.
5. Cook the noodles according to the packet instructions and add them to the pot.
6. Finally, add the lime juice and simmer over a low heat for 15 minutes.
7. Stir in the remaining coriander and dish up, garnished with sesame seeds.

Superfood Buckwheat and Thyme Skin-soothing Salad

The ingredients in this superfood salad are brimming with tasty, targeted vitamins and minerals to promote optimum skin-microbiome functioning. The cocktail of skin-nourishing benefits includes: vitamin A for healing, B vitamins, vitamin C and zinc. An underused skin-health hero, the small but mighty sprouted grains are quite literally loaded with a powerhouse of amino acids and phytochemicals that work together with the rest of the wholefood ingredients in this recipe to promote the regeneration of skin cells, encourage collagen synthesis and support healing.

Serves 2

1 butternut squash (or use sweet potatoes), peeled and cut into small chunks
1 tbsp extra-virgin olive oil
1 tsp dried thyme
Pinch each of sea salt and freshly ground black pepper
200g buckwheat
100g broccoli, cut into small chunks
½ cucumber, chopped
1 Romano pepper, chopped
30g (approximately) parsley or dandelion leaves, finely chopped
Large handful of rocket
1 avocado
Handful of sprouted grains (alfalfa, mung bean, broccoli or a mix)
2 tbsp pumpkin seeds
2 tbsp finely chopped almonds

For the skin-healing dressing

1 garlic clove, grated
Juice of ½ lemon
4 tbsp extra-virgin olive oil
1 tsp raw honey
1 tbsp apple cider vinegar
1 tsp Dijon mustard
½ tsp dried thyme
Pinch each of sea salt and freshly ground black pepper

1. Preheat the oven to 190°C.
2. Place the butternut squash in a roasting tin, drizzle with the extra-virgin olive oil and season with the thyme, sea salt and pepper. Roast for 35 minutes, or until perfectly golden and slightly browned around the edges.
3. In the meantime, cook the buckwheat according to the packet instructions.
4. Combine all the ingredients for the dressing.
5. Construct your salad by adding the roasted butternut squash, buckwheat, broccoli, cucumber, pepper, parsley or dandelion leaves and rocket to a bowl. Chop the avocado and add to the bowl, then top with the sprouted grains, pumpkin seeds and chopped almonds.

Restorative Beetroot Soup

Sweet, creamy and with real depth of flavour and a fiery edge, this restorative, vitamin-rich soup is bursting with phytonutrients to nourish your skin and counter skin inflammation presenting as breakouts. This is because beetroot contains nutrients called betalains that boost the detoxification work of the liver and, as you know, optimal liver function is vital for clear and healthy skin.

It tastes amazing, works a treat in helping you feel instantly energised and it freezes well, too. So next time you're in need of some quick and easy nourishment, you'll have a soup ready to go!

Serves 3–4

500g steamed beetroot, halved
3 small sweet potatoes, chopped into small chunks
1 red onion, roughly chopped
1 tbsp extra-virgin olive oil
1 tsp smoked paprika
2 tsp ground cumin
½ tsp ground coriander
½ tsp cumin seeds
Pinch each of dried chilli flakes, sea salt and freshly ground black pepper
1 tbsp raw honey
400ml full-fat coconut milk
7½cm piece of fresh ginger, grated
4 garlic cloves, grated
400ml vegetable stock (made with 1 cube)

1. Preheat the oven to 180°C.
2. Place the beetroot, sweet potatoes and red onion in a lined roasting tin. Drizzle with the extra-virgin olive oil, sprinkle with the paprika, ground cumin and coriander, cumin seeds, chilli flakes, salt, pepper and honey and pop in the oven for 35 minutes, or until the vegetables are soft, slightly browned around the edges and gorgeously caramelised.
3. In the meantime, pour the coconut milk into a saucepan over a medium heat, add the ginger and garlic and stir. Let it simmer for a few minutes, then remove from the heat and set aside.
4. Pour the stock into a blender, along with the roasted vegetables and coconut milk and blitz until smooth. Check the consistency and if you want it less thick, add some boiling water to loosen it up. Serve on its own, or you could stir in some quinoa for extra plant protein or dip in some seeded crackers.

Kefir Blueberry Breakfast Smoothie

This truly delicious blueberry breakfast smoothie is my absolute go-to during the spring and summer months of the year and my cycle, too! It's a great way to get a big hit of hormone-nourishing nutrients, including antioxidants, probiotics, protein and even apoptogenic mushrooms to keep you energised, focused and satisfied until lunchtime.

Not only is this a great breakfast option, it's also one of my favourite ways to nourish my skin. That's because it has a good base of blueberries – an undeniable skin superfood. These small but mighty berries are loaded with flavonoids that act as a powerful antioxidant in the body to keep inflammation at bay and calm angry skin from within. They are beneficial for all skins and especially so if you suffer with rosacea, as they're known to strengthen damaged blood vessels to help improve broken capillaries. And as if all that isn't impressive enough, blueberries are a rich source of vitamin C, which boosts collagen and supports skin elasticity.

Serves 1

250ml kefir
Handful of frozen blueberries
Handful of frozen mango
1 tbsp frozen cauliflower rice or 1 frozen cauliflower floret (optional)
1 serving of unflavoured quality plant-protein powder
2 tbsp seed-cycling seeds (according to your cycle phase, see pages 40–43)
1 tbsp nut butter
1 tsp tremella medicinal mushroom powder to support skin (or try lion's mane powder for increased focus and mental clarity)

1. Pour the kefir into a blender, followed by the remaining ingredients and blitz until smooth.

TIP: Adding cauliflower to your smoothies is a great way to boost it with B vitamins, vitamin C and insoluble fibre to prevent constipation and keep your bowels moving (vital for clear and healthy skin). It also contains a plant compound called indole-3-carbinol (I3C), which acts as a plant oestrogen to balance hormones by regulating oestrogen levels. The added creaminess from frozen cauliflower is an added bonus for your taste buds!

Immunity Tea Cubes, Two Ways

Increasing your hydration is a low-hanging fruit when it comes to helping your hormones and optimising your overall health and wellbeing. And drinking grated ginger and lemon first thing in the morning is a wonderful way to start your day as you mean to go on. With hydration at the forefront, this will help every system in your body to function better, including its ability to detoxify and encourage your stool to pass more easily along your digestive tract.

We lose hydration as we sleep, so rehydrating as soon as you wake up should be a non-negotiable, while adding in nature's anti-inflammatories is an ideal way to get the most out of your morning cuppa – supporting your gut, aiding liver function, helping to tackle bloating, indigestion and pregnancy nausea and just to show your immune system some TLC.[95]

Lemon has alkalising properties and helps to balance the body's pH levels. It also contains limonene that helps digestion by moving food along your digestive tract.[96] Studies show that ginger has a pro-kinetic effect, which promotes movement in the gastrointestinal tract to help keep you regular, reduces inflammation and tackles bloating.[97] Ginger also improves bile acid, which leads to pancreatic enzyme production, helping to improve your digestion and protect the gastric mucosa for a healthy gut. Turmeric is a potent anti-inflammatory that can be made more bioavailable in the presence of black pepper, and the ground cinnamon (if used) encourages stable blood-sugar levels.

You'll need an ice-cube or baby-food tray and gloves (to protect your hands from turmeric stains). And please note, you can either make this using just lemon and ginger and add in the turmeric, black pepper, cinnamon and raw honey if you feel run down and in need of some immune support.

Makes 9 large cubes

For Immunity Tea Daily
500ml filtered water
4 lemons
7½cm piece of fresh ginger, grated

For Immunity Tea Boost
500ml filtered water
4 lemons
5cm piece of fresh ginger, grated
5cm piece of turmeric root, grated
Generous pinch of freshly ground black pepper
½ tsp ground cinnamon
1 tsp raw local honey, to serve

1. Pour the filtered water into a jug, then halve the lemons and juice them straight into the jug.
2. Add the ginger to the jug (or, for the Immunity Tea Boost, add the ginger, turmeric, black pepper, honey and cinnamon – if using).
3. Give your immunity water a good stir and pour into your ice-cube or baby-food tray (I put a plate underneath the tray to avoid spillages, as turmeric stains and can ruin your worktops).
4. Pop in the freezer.
5. Once frozen, add a cube into a cup, pour boiling water over it stir and drink (adding some raw honey for its medicinal properties if you are feeling run down or fancy a subtle sweetness).

CYCLE SYNC | FOLLICULAR/OVULATORY

Creamy Greens, Happy-hormone Smoothie

Another spring/summer smoothie option that's rich in nutrients, including anti-inflammatory chlorophyll, energy-boosting B vitamins, calming magnesium, thyroid-supportive selenium and plant protein to keep you sustained. A great way to get your greens into your system first thing to show your hormones you love them.

Serves 1

250ml unsweetened nut milk
Handful of spinach
3 frozen cauliflower florets or cauliflower rice
½ banana
Handful of frozen mango
1 tbsp hemp seeds
1 tbsp nut butter
1 Brazil nut
Seed-cycling seeds (see pages 40–43), as appropriate

1. Add all the ingredients to a blender and blitz until smooth. Simple as that!

Tip: If you do decide to drink smoothies outside out of your follicular or ovulatory phases, let them come to room temperature first. Your body needs warm nourishment during this time, not harsh cold drinks, which can amplify PMS in some women (thought to be due to the release of prostaglandins).

CYCLE SYNC | LUTEAL/MENSTRUAL

'A Cup of Calm' Cosy Golden Milk

This incredibly cosy, immune supportive golden milk recipe is inspired by Auyverdic traditions. The active ingredient within turmeric, curcumin, gives turmeric its distinctive yellow colour and contains powerful antioxidants known for promoting healthy ageing and proper liver function. Whenever I drink a cup of golden milk it reminds me of my best friend, Sarita, not just because it's so warming like her, but because she told me about the benefits of turmeric before I even understood what it was all those years ago and now I can't get enough of it!

Serves 1

½ tsp organic ground turmeric
1/4 tsp ground cinnamon
Pinch of black pepper to activate the curcumin in the turmeric
200ml warm milk (unsweetened if using plant milk), preferably in a milk frother for a dreamy consistency
Dash of raw honey or maple syrup

1. Simply add the turmeric, cinnamon, black pepper and raw honey to a mug with warm milk, mix well and drink slowly, enjoying every nourishing sip.

TIP: Drink in the evening to help stimulate 'rest and digest' and activate the parasympathetic nervous system for a cup of calm before bedtime..

For an added benefit, add collagen powder for an even bigger nutrient hit.

Every-season Tomato, Fennel and Buckwheat Soup

Everyone needs a reliable, classic tomato-soup recipe and this one is even better because it can be tailored to your cycle: just serve it hot during your luteal and menstrual phases and cold during the follicular and ovulatory phases. Then soak up the benefits of the lycopene-rich roasted tomatoes (even more so when they are cooked) with their many anti-inflammatory properties to support your heart, protect your skin and reduce oxidation in the body.

This soup keeps in the fridge for four days and freezes well, too.

Serves 2

1 fennel bulb
2 red onions, chopped
2 red bell or Romano peppers, chopped
300g cherry tomatoes, halved
1 garlic bulb, ends chopped off
1 tsp fennel seeds
1 tbsp Italian seasoning
Pinch of dried chilli flakes
Pinch each of sea salt and freshly ground black pepper
1 tbsp extra-virgin olive oil
150g precooked buckwheat (or swap for quinoa)
30g basil
1 vegetable-stock cube, dissolved in 400ml water
2 tbsp tomato purée
1 tbsp balsamic vinegar
Mixed seeds, to serve

1. Preheat the oven to 190°C.
2. Chop the ends off the fennel, cut the bulb in half and cut out the core. Slice the rest of the fennel.
3. Transfer the vegetables to a large roasting tin, along with the garlic bulb, and sprinkle with the fennel seeds, Italian seasoning, chilli flakes, salt and pepper and drizzle with extra-virgin olive oil.
4. Roast in the oven for 45 minutes, tossing the vegetables halfway through cooking time.
5. Meanwhile, cook the buckwheat or quinoa according to the packet instructions.
6. Remove the vegetables from the oven and transfer to a blender or food processor along with the basil. Squeeze in the garlic (discarding the skins).
7. Pour in the vegetable stock, add the tomato purée and balsamic vinegar and blitz until smooth, adding more water to loosen the consistency if needed.
8. Stir in the buckwheat and serve, sprinkled with mixed seeds for extra nutrients and plant points.

TIP: This soup doubles up as a delicious tomato-based pasta sauce. SImply follow the guidance above and skip the added grains!

7. The Neuroendocrine System

'A measure of true success is a calm nervous system.'

Made up of billions of nerve cells, the nervous system is your internal communication centre, controlling all your body's functions via messages between brain and body. It operates very closely with your endocrine system in an incredibly intimate relationship, continuously working to maintain balance physically and mentally, being responsible for metabolism (impacting weight), reproduction, emotions, mood, sleep and hormone production.

These systems help to regulate the electrical and chemical processes that keep your hormones working and in balance.[98] They are linked by the hypothalamus (imagine it being a bridge between the two), which is a tiny collection of nuclei at the base of the forebrain that controls an astonishing amount of human behaviour, including emotions and how you respond to stress. This combination of systems is known as your 'neuroendocrine system'.

What's This Got to Do with Hormone Balance?

The health of your neurotransmitters influences the internal terrain in which your hormones are produced and how well they communicate with each other, keeping you feeling calm, safe, happy, energised, motivated and emotionally and mentally stable throughout your life.

It's a two-way street: the brain produces and responds to hormones circulating in your blood, but when there is a hormonal imbalance in the body, it automatically affects your nervous system. When brain health is compromised it can negatively impact your hormone levels by impairing the normal production, secretion and signalling processes of the many neurotransmitters needed for good health.

If this communication is not operating as it should, it will affect the delicate orchestra of hormones. Hence, the close relationship between your nervous and endocrine system is vital to achieving hormonal harmony.

So if you are experiencing low mood, irritability, anxiety, brain fog, fatigue and burnout, struggling with sleep or going through peri- and menopausal changes impacting these areas of your cognitive, mental and emotional health, this chapter is for you.[99] And you will be surprised just how much you can support this integral system via the food you nourish your body with and the grounding daily habits you start to cement into your life.

What Does the Neuroendocrine System Do?

The neuroendocrine system is responsible for the hypothalamic control of the pituitary hormones, ensuring they respond properly and 'normally' to environmental stimulation through regulated secretion of hormones and neurotransmitters.[100]

The hypothalamus is constantly interpreting these signals from the environment and sending hormonal alerts to the pituitary gland (located at the base of the brain), which then sends hormonal signals to other endocrine glands to ensure a smoothly functioning system and to promote overall hormonal balance, helping you to feel stable from a cognitive perspective.

Pituitary hormones include the following:

Thyroid-stimulating hormone (TSH) TSH signals the thyroid gland (located in your neck) to produce and release thyroid hormones, which play a major role in regulating the body's metabolism (the process that turns food into energy). These hormones are called triiodothyronine (T3) and thyroxine (T4). It's common to have thyroid disorders (much like I did), where the body produces too much or too little thyroxine, and this can impact the delicate balance of your hormonal system. TSH is the standard test for thyroid dysfunction.

Follicle-stimulating hormone (FSH) and luteinising hormone (LH) FSH and LH communicate with the ovaries, promote ovulation and the production of oestrogen, testosterone and progesterone.

Adrenocorticotropic hormone (ACTH) ACTH is released from the pituitary in response to stress, and signals to the adrenal glands to produce the stress hormone cortisol.

Prolactin Prolactin stimulates breast-milk production and is responsible for the development of mammary glands within breast tissues.

Antidiuretic hormone (ADH) ADH, also called vasopressin, regulates sodium and fluids, affecting blood pressure.

MELATONIN: YOUR MASTER SLEEP HORMONE

The pineal gland in the brain is responsible for melatonin production, which is stimulated by darkness and promotes feelings of sleepiness, regulates your sleep-wake cycle, normalises your circadian rhythm (your internal body clock) and influences the health of the reproductive and immune systems.[101] Adequate melatonin levels ensure that you get the restorative sleep needed for a balanced hormone system and, therefore, good energy and consistent mood.

How to get more melatonin

Melatonin comes from serotonin, 95 per cent of which is produced in your gut microbiome.[102] Serotonin is made from the amino acid tryptophan, which also plays a crucial role in many other metabolic functions, including the modulation of the endocrine system and the impact of cortisol, prolactin and growth hormone, which impacts everything from your cycle health and length, focus and how stressed you feel.[103]

Sunlight exposure in the morning resets your internal body clock, improves your sleep–wake cycle and plays a significant role in regulating your mood, reducing anxiety, depression and insomnia.[104] This is because sunlight exposure in the morning naturally boosts serotonin levels, and, as we've discovered just now, you need serotonin to make your master sleep hormone, melatonin. Without natural light exposure, it's harder for your brain and body to know when it's time to go to sleep at night and when to naturally wake you up in the morning. And the same goes for blue-light exposure from screens at the wrong times (i.e. at nighttime), as this prevents melatonin production and stops you feeling sleepy.

TIP: Open your curtains/blinds as soon as you wake up to flood your room with natural light, then get straight outside into your garden or balcony, if you have one, or just stick your head out of the window to expose your naked eyes. Serotonin is produced via retina exposure to natural light, so avoid sunglasses in the morning for the first five to ten minutes, otherwise you will stop this magic process.

Why (Complex) Carbs at Dinnertime Will Reduce Stress and Help You Sleep

If you are stressed, your body is producing cortisol. Progesterone is a precursor to cortisol, so if you are under chronic stress, you might find your progesterone is depleted, which can mean there is too much oestrogen (when there is not enough progesterone to balance this out). Eating complex carbs (such as brown rice, sweet potato, buckwheat and quinoa) in your evening meal supports your sleep hormones by blunting cortisol, raising serotonin and increasing levels of GABA (that anti-anxiety neurotransmitter). I used to skip carbs at dinner to cut calories, but I'd spend all evening fighting the urge to snack on sugar because I was unsatisfied and would be tossing and turning all night and never feeling properly rested. I wish I had known how important nature's carbs were in my healing and reblancing journey.

HOW YOU HANDLE (OR DON'T HANDLE) STRESS

Stress can trigger your body's response to a perceived threat or danger (known as the fight-or-flight response), when hormones like adrenaline and cortisol are released on auto, speeding up your heart rate, causing your palms to become sweaty, slowing down your digestion and making your muscles feel tense.

Does this feeling resonate with you? It certainly does for me, having robbed me of some unbelievable opportunities in my career. This was because for so long I wasn't taking care of myself and didn't understand the importance of prioritising the health of my neuroendocrine system.

My abnormal response to stress caused me to feel, at best, anxious and discontented and, at worst, caused panic attacks that made me feel like I was having a heart attack. It got to the stage where I feared social situations and even walking down my local high street.

I now know these abnormal responses were, in fact, driven by my consistently poor diet and lifestyle choices, and it was only when I started nourishing my body and my mind and put my health and wellbeing first that these symptoms stopped. Nowadays, I occasionally feel anxious in the lead-up to my period, but I have all the tools – via my nutrition and by honouring rest and healing time – to be able to care for myself, instead of going against my neuroendocrine system and allowing it to become a problem.

For example, I was undernourishing by restricting my meals in my quest to be 'thin', and therefore limiting my essential nutrient intake for my brain and hormones to be able to function properly. I would feel weak, sluggish and spend a lot of the day in 'hangry' mode (poor James!). I would also use caffeine, sugar and sometimes alcohol to – ironically – make me 'feel better' and handle my stress. I became borderline obsessed with intensely working out most days (I didn't have a cycle at the time, but I know I wouldn't have been in the mindset to work out in alignment with it even if I had), which was causing my cortisol levels to be constantly spiked (you can literally feel this happening in your body if you tune in). And then, to top things off, I stayed up until gone midnight every night watching TV or scrolling on my phone. I was living in my sympathetic nervous system (fight or flight) and my poor parasympathetic nervous system never got a look in.

How Your Cycle Impacts Your Mood

Multiple factors contribute to low mood (aka low serotonin) and because of that your energy, too, some of which you can't control or predict. But there are many internal processes you absolutely can control via functional food intervention.

Your mood and energy levels (and how you handle the inevitable stresses of life) are, in fact, heavily influenced by your menstrual cycle and how optimised it is. Therefore, your mood hormones such as serotonin, dopamine, noradrenaline, oxytocin and cortisol are heavily impacted by fluctuating oestrogen. In the case of peri-/menopause, the decline of your cycle and the impact of your dipping hormones on your brain might manifest as a range of symptoms, including low self-esteem, feeling mentally unstable and unhappy/discontented but without a specific cause. Supporting the decline of your cycle and the balance of your sex hormones at this time will impact your nervous system/brain health and, therefore, how happy, content and energised you feel through the ebbs and flows of your hormonal life.

If you have a cycle, have you noticed how short your fuse is in the lead-up to your period during your luteal phase? And perhaps you feel better during the follicular phase of your cycle, after your period? Well, this is driven by the balance of hormonal chemical messengers and neurotransmitters circulating your body that enable you to feel happy, energised, motivated and stable.

As we've already learned, when these are out of whack and haven't been given or, more accurately, 'fed' the right supply of nutrients from your diet, along with adequate movement and sufficient sleep, your mood and how you are able to show up in the world will be impacted.

This plays a significant part in your relationships, potentially how you parent and how motivated, driven and fulfilled you feel at work, and is why I cannot stress enough (pardon the pun) that taking care of yourself by investing in your health and wellbeing and keeping your cup filled up is not selfish or self-centred, but is essential to your mental, emotional and physical health.

THE VAGUS NERVE: FROM FIGHT OR FLIGHT TO CALM

The vagus nerve is the largest nerve in your body, running from the base of your brain and down through the neck, where it branches out into the chest and torso. It plays a big role in regulating your breathing, heart rate, inflammation, muscles, digestion, circulation, emotions, metabolism and stress hormones.

The good news is, you can easily activate your vagus nerve to give your neuroendocrine system extra support by reducing cortisol and stimulating 'rest-and-digest' mode; you will instantly feel the positive impacts, as calming, nourishing energy flows through your cells and tissues, helping to heal you from within. Try the following:

- The Nourish Method Nervous System Reset Ritual (see page 185)
- A cold shower with the water directed at the back of your neck
- Singing, laughing or humming
- Cat–cow stretch
- Supporting your gut health; the vagus nerve connects the brain to the gut, and vice versa

YOUR MOOD-HORMONE GLOSSARY

Dopamine, your 'feelgood' reward hormone is a neurotransmitter that's an important part of your brain's reward system. It's associated with pleasurable sensations, along with learning, memory and more.

Noradrenaline, your 'energy' hormone (also called norepinephrine) is the primary neurotransmitter in the sympathetic nervous system in the brain, where it works to control energy levels, heart rate, blood pressure, liver and other functions.

Serotonin, your 'happy' hormone helps to regulate your mood as well as your sleep, appetite, digestion, learning ability and memory.

Oxytocin, your 'love' hormone is essential for childbirth, breastfeeding and strong parent–child bonding. It can also help promote trust, empathy and bonding in relationships. Levels generally increase with physical affection.

Endorphins, your 'pain-relief' hormones are your body's natural pain relievers, produced in response to stress or discomfort. Levels may also increase when you engage in reward-producing activities, such as eating, working out or having sex.

Cortisol, your 'stress' hormone is often misunderstood and seen as something to fear, but we do need it. It gets us out of bed in the morning and motivates us in life, but it's all about keeping it regulated, healthy and in balance.[105] High cortisol can lead to anxiety, high blood pressure, weight gain, inflammation and being unable to properly deal with stress, while low cortisol can cause symptoms of fatigue, loss of appetite and weight loss. Elevated cortisol can also adversely affect hormone balance, eventually downregulating the production of sex hormones, such as progesterone, which can cause irregular cycles and even loss of periods (amenorrhea). The good news is, increasing hormones such as oxytocin, by spending time with people who light you up and by prioritising self-care time to relax or do activities that trigger endorphins and lift your mood, can blunt the negative impact of too much cortisol.

WHICH HORMONES IMPACT AND DICTATE YOUR MOOD?

Understanding which hormones are responsible for *why* you feel the way you do is helpful because it enables you to eat in a way that supports them. And by working with your hypothalamus in this way to regulate nervous-system health, you will be proactively improving how you feel in body and mind.

Some of the key hormones that impact you and need to be in balance are oestrogen, progesterone, insulin, cortisol and thyroid. They cross the blood–brain barrier (the protective interface between the central nervous system and the circulating blood[106]) and bind to receptors on neurons (which are brain cells).[107] When the brain experiences drops in oestrogen and progesterone due to generalised hormone imbalance, a condition such as PCOS or, during perimenopause and menopause, symptoms such as dips in mood, brain fog, and hot flushes can occur.[108] Be assured, a healthy brain has the ability to learn new things and adapt to a new environment, even in your later years, and brain-/nervous-system-related symptoms of perimenopause tend to subside once the brain adjusts to the new hormonal landscape of menopause, supported by a healthy brain-nourishing diet.[109]

OESTROGEN AND MOOD

It was only a matter of time before the Queen Bee hormone showed up again! Oestrogen really deserves its own chapter, but to keep it simple and to the point: your mood is heavily dictated by the fluctuation of oestrogen throughout your monthly cycle and this effect is often exacerbated during the menopause transition. Changing oestrogen levels can result in mood swings and can impact self-confidence, making you more prone to anxiety and depression.

Women are naturally at a much higher risk than men of developing mood disorders and depression, which both normalises mood swings but also give us the motivation to help our bodies cope with this fluctuation of oestrogen more efficiently, because knowledge is power.[110]

Eating To Support Brain Health and Hormonal Balance

Here, we're going to take a look at some of the key functional food sources that will help your brain to function and thrive, improving mental clarity, drive and focus and helping to naturally banish the brain fog, low mood and anxiety associated with hormone imbalance. Hormone imbalance affects the brain and its ability to carry out its job properly. At the same time, when brain health is compromised, it affects your hormone levels. So brain health is essential for both hormones entering the brain and the hormone signals leaving the brain and for keeping everything running smoothly, for the maintenance of internal homeostasis. A healthy brain and hormone balance really do go hand in hand.

But as well as looking at brain-nourishing foods to add to your diet, it's important to mention those that will hinder brain health, including, unsurprisingly, processed foods, refined sugars, emulsifiers, alcohol and – a big one – highly refined, low-quality vegetable oils.

HEALTHY FATS/OMEGA-3 FATTY ACIDS

Your brain needs healthy fats to function. Think of them as fuel for your brain helping to transport nutrients around your body.

Studies show that foods rich in EPA and DHA (types of omega-3 fatty acids) have a direct impact on mood and are associated with the prevention of depression.[111] Include oily fish in your diet, using the acronym SMASH to remember which ones contain the highest amount of omega 3: **S**almon, **M**ackerel, **A**nchovies, **S**ardines and **H**erring.

Plant-based healthy fat sources (ALA) include nuts, seeds and olive oil and should be eaten as part of a daily, healthy, balanced diet.

CARBOHYDRATES

Carbohydrates are glucose, which is energy for our cells and the preferred fuel source for our brains.

Studies show that consumption of complex carbohydrates correlates with successful brain ageing and improved memory, both in the short and long term, as opposed to an intake of simple carbohydrates (white, refined), which is consistently associated with decreased global cognition.[112]

The bottom line? Carbs don't need to be feared and are absolutely essential when it comes to hormone balance and brain health. Just be smart with the carbs you include in your diet, focusing on high-fibre complex carbs grown from nature that take longer to digest and release energy slowly (such as wholegrains, quinoa, brown rice, buckwheat and sweet potatoes). They are richer in nutrients than refined simple carbs, which will spike your blood-sugar levels quickly and are associated with weight gain and poor metabolic health.

Also, low-carb or keto diets affect female hormones and the menstrual cycle. Insufficient starch intake can suppress luteinising hormone and ovulation and we can't balance hormones without this, so don't demonise carbs. (And if you have done so up until this point, your hormones will be delighted if you bring them back.)

GUT-FEEDING FERMENTED FOODS

Your gut can influence the production of some key neurotransmitters, including GABA and serotonin, and a healthy gut microbiome equals healthy serotonin levels, resulting in better moods. Simple. Support your neurotransmitters by increasing the fibre in your diet, using the high-fibre recipes throughout this book (especially in this chapter and the gut chapter – see pages 65–109) and by being proactive with adding fermented probiotic-rich foods into your diet (including kimchi, kefir, sauerkraut, live yoghurt and miso) and prebiotics from foods such as garlic, onions, asparagus, Jerusalem artichokes, mushrooms, leeks, parsnips, almonds, apples and cacao.[113,114,115]

MAGNESIUM – NATURE'S TRANQUILLISER

Otherwise known as nature's tranquilliser, magnesium soothes your nervous system and is a powerful stress reliever. It regulates the HPA axis, promotes sleep, plus it activates our GABA receptors, which calm us down. It's a miracle mineral for periods, as it supports insulin and thyroid hormone and the healthy metabolism of oestrogen, which is why it's so essential when it comes to overall hormone balance.

Food sources include leafy greens and nuts and seeds, but stress easily depletes magnesium stores, making lots of people deficient. Supplementing is therefore helpful but consult with your nutritionist to ascertain which variety you need (magnesium glycinate is a great all-rounder and is gentle on the gut).

YOUR AT-A-GLANCE GUIDE TO BRAIN FOOD

Blueberries are high in polyphenols – antioxidants that are beneficial for gut bacteria and cognitive function.

Broccoli contains glucosinolates, which, when broken down by the body, produce isothiocyanates, molecules that neutralise carcinogenic toxins in the body and help lower the risk of neurodegenerative diseases, such as Alzheimer's and Parkinson's.

Leafy greens (kale, spinach, rocket) are rich in essential B vitamins, and vitamins E and K, along with an abundance of minerals that support brain health and improve memory and focus.

Pumpkin and sunflower seeds are packed with antioxidants and minerals, including magnesium, zinc, iron and copper, which help to protect the brain from free-radical damage and facilitate effective nerve signalling. Sunflower seeds are also rich in thiamine (B1), an important vitamin for memory and cognition.

Brazil nuts contain ellagic acid (a type of antioxidant) that has anti-inflammatory properties to protect the brain from damage. They are also high in the mineral selenium, which is needed for brain-signalling pathways.

Note: nuts and seeds are best soaked overnight for at least seven hours, as this activates enzymes to aid absorption and enable better digestion. Discard the water afterwards.

Bananas are rich in the amino acid tryptophan, which converts to serotonin in the body to help support healthy gut function and mood.

Turmeric contains curcumin, a powerful compound with anti-inflammatory and antioxidant properties, which has been shown to cross the blood–brain barrier to help brain cells grow, reduce mental decline and improve memory and mood.

Wholegrains including oats, quinoa, buckwheat, barley and millet contain an abundance of B vitamins and other minerals to reduce inflammation in the brain and improve cognitive function.

Beans and pulses (lentils, chickpeas, kidney beans) are high in fibre, B vitamins and omega-3s. They provide a steady supply of fuel (glucose) to the brain, which aids concentration and memory.

ANTIOXIDANT-RICH RAINBOW FOODS

The science suggests that antioxidants may benefit mental health by reducing inflammation and oxidative stress in the brain that can lead to depression symptoms.[116]

Free radicals, as we saw earlier (see page 70), are molecules in your body that are unstable and can cause cell damage, contributing to hormonal imbalances. But the good news is that antioxidants can stabilise free radicals.

Antioxidant-rich foods include all bright, colourful vegetables (such as beetroot, aubergine, sweet potatoes and red cabbage), fruits (such as all types of berries), nuts (such as Brazil nuts) and seeds (including pumpkin), herbs and spices (such as turmeric, ginger, garlic, rosemary, parsley and sage), wholegrains (such as buckwheat and black rice), legumes (such as black beans), healthy-fat sources (like extra-virgin olive oil and avocados) and not forgetting green tea and dark chocolate, too.[117]

Minimise Your Exposure to UPFs

Different foods have the power to promote health and energise the mitochondrial cells in your body and brain (your battery pack, remember?) or to drain your energy stores and deplete your health and wellbeing. Ultra-processed foods (UPFs) sit firmly in the latter camp. In fact, there is evidence strongly linking UPFs with a higher risk of depression and anxiety.[118]

I could talk about this subject until the cows come home, but the bottom line is: UPFs have zero nutritional and hormonal health benefits. So if you want to better your physical and mental health and promote hormone balance by working *with* your body (and stop working *against* it by fuelling it with unwanted chemicals), you need to minimise these offenders in your diet. And you know the ones ... I don't need to list them here.

The majority of foods you eat should be wholefoods, grown from nature, and avoid buying ultra-processed foods laden with inflammatory refined vegetable oils, sugars and ingredients you can't even pronounce. Think more real foods and minimal fake foods and you will be on the road to better hormonal health with every meal you consume.

To conclude, never underestimate the critical role that food plays in your mental, emotional and physical health – because the way you support your body's neuroendocrine system, using food and lifestyle as your daily medicine, will help to normalise your body's neurotransmitters, helping you to handle daily stressors like anxiety and improving your mood and energy throughout your cycle and beyond.

The Nourish Method Nervous System Reset Ritual

This ritual is so simple, takes almost no time, requires no equipment and can be done anywhere and anytime you need to reset and realign yourself. It's the ultimate nervous-system tranquilliser and something I always come back to. Although you can do the ritual in any position, try to sit with your back straight to get started. Place the tip of your tongue against the ridge of tissue just behind your upper front teeth and keep it there throughout the entire ritual. You will be exhaling through your mouth around your tongue; try pursing your lips slightly if this seems awkward.

1. Exhale completely through your mouth, making a whoosh sound.

2. Close your mouth and inhale quietly through your nose to a mental count of 4.

3. Pause.

4. Exhale completely through your mouth, making a whoosh sound to a count of 7.

5. Now inhale again and repeat the cycle 3 more times.

Note that you always inhale quietly through your nose and exhale audibly through your mouth. The tip of your tongue stays in position the whole time. Exhalation takes twice as long as inhalation. The time you spend on each phase is not important; it's the ratio of 4:7 that matters. With practice, you will get used to inhaling and exhaling more and more deeply and will feel reset each time you do it.

Unlike tranquilising drugs, which are often effective when you first take them but then lose their power over time, this ritual is subtle when you first try it, but gains in power with repetition and practice. Do it at least once a day and increase as needed when life feels overwhelming and you need to regulate your nervous system. It's a useful free tool you will always have with you.

Another nervous system soothing ritual I love to do when I have more time is to submerge my face into a bowl of ice-cold water. This might sound odd, but it helps relieve anxiety by activating the body's mammalian diving reflex, which sends a calming message to the vagus nerve, stimulating your parasympathetic nervous system to slow down your heart rate and breathing – essentially stopping the 'fight or flight' response in its tracks and activating 'rest and digest mode', leaving you feeling calm, grounded and reset in body and mind. Simply start by filling a bowl (ideally one that can fit your entire face) with water and ice. Then submerge your face into the iced water for 10–15 seconds. Repeat as needed. Most importantly, be gentle with yourself.

Neuroendocrine System Recipes

Cajun Maple Chicken, Butter Beans and Greens

Lean protein like chicken supports hormone balance and egg maturation by providing essential amino acids, the building blocks of female hormones that maintain reproductive health and keep your cycle regular.

In this recipe, the chicken is combined with the high-fibre butter beans to keep you regular and everything moving in the right direction via your bowels; and they also serve as an alternative complex carbohydrate for sustained energy. Balanced with magnesium-rich leafy greens to replenish depleted stores, this meal is the ideal weeknight one-tray, it tastes good and is also good fuel for your whole body.

Serves 2 (plus a little one!)

500g organic chicken thighs
1 red onion, roughly chopped
1 red pepper, roughly chopped
Handful of cherry tomatoes, roughly chopped
200g tenderstem broccoli, roughly chopped
4 garlic cloves, roughly chopped
2 x 400g tins butter beans, drained and rinsed
2 tsp Cajun seasoning
1 tsp thyme
1 tsp garlic granules
Pinch each of sea salt and freshly ground black pepper
Drizzle of extra-virgin olive oil
Drizzle of maple syrup
2 handfuls of leafy greens

For the marinade

2 tbsp Cajun seasoning
1 tsp thyme
1 garlic clove, grated
1 tsp mild chilli powder
3 tbsp extra-virgin olive oil
Pinch each of sea salt and freshly ground black pepper
1 tsp maple syrup

1. Preheat the oven to 200°C.
2. Prepare the marinade by combining the ingredients together in a bowl.
3. Place the chicken thighs in a bowl, pour in the marinade and toss the thighs in it. Set aside for at least 15 minutes (or longer, if you can) to soak up the flavours!
4. In a big roasting tin, add the onion, pepper, tomatoes, tenderstem and garlic cloves. Pat dry the butter beans and throw them into the tray, too.
5. Sprinkle with the Cajun seasoning, thyme, garlic granules and a pinch sea salt and pepper. Finally, drizzle with extra-virgin olive oil and a little maple syrup.
6. Add the chicken thighs and roast in the oven for 50 minutes, until slightly browned, turning halfway through the cooking time.
7. Serve with leafy greens and if you're extra hungry during your menstrual phase, add some wholegrains like buckwheat, too.

My Replenishing Classic Chilli

Rich in haem iron, the most bioavailable iron source, my classic chili recipe will replenish mineral stores lost through your menstrual blood to help restore you when you need it most. Enriched with vitamin C to aid iron absorption, from the lycopene-loaded tomatoes, this nourishing, warming bowl of flavour and fibre is the definition of menstrual-phase nutrition, ticking all the nutritional boxes and more. Serve with a side of wholegrain brown rice and soak up every restorative mouthful.

Serves 4

Drizzle of extra-virgin olive oil
1 large red onion, finely chopped
2 red peppers, finely chopped
4 garlic cloves, grated
Pinch each of sea salt and freshly ground black pepper
1 heaped tbsp smoked paprika
1 heaped tbsp ground cumin
1 tbsp mild chilli powder
2 tbsp balsamic vinegar
400g grass-fed beef mince
400g tin chopped tomatoes
2 tbsp tomato purée
1 tsp Worcester sauce
1 tbsp raw honey
1 tsp dried thyme
1 tsp fennel seeds
1 tsp dried rosemary
1 beef stock cube
2 fresh chillies, finely chopped, or 1 tsp dried chilli flakes
400g tin kidney beans, drained and rinsed
400g tin black beans, drained and rinsed
2 squares of dark chocolate (70% cocoa solids)
1 bay leaf
250g brown rice
Classic Guacamole (see page 59)
Juice of 1 lime
Bunch of fresh coriander, roughly chopped

1. Heat the extra-virgin olive oil in a large pot over a medium heat. Add the onion, peppers and garlic, along with the sea salt and pepper. Cook for around 5–10 minutes, until soft.
2. Add the paprika, cumin and chilli powder, combine and cook for 1 minute. Pour in the balsamic vinegar and allow the mixture to caramelise for 1 minute.
3. Add the mince, breaking it up with a wooden spoon, then cook for 5 minutes or until the meat has browned, before adding the tomatoes, tomato purée, Worcester sauce, honey, thyme, fennel seeds, rosemary, stock cube and chillies.
4. Shake any excess water off the beans, using a colander, then add them to the pot, along with the chocolate and bay leaf. Stir and cover the pot with a lid. Reduce the heat to low and simmer for 1 hour (or 2, if you have time).
5. Meanwhile, cook the brown rice according to the packet instructions and make the guacamole following the instructions on page 59.
6. Before serving, squeeze the lime juice into the pot and stir in the coriander.

'Make-it-your own' Simple Salmon Traybake

Having this simple salmon traybake in my bank of go-to hormone-balancing recipes has made it a weekly ritual, and I make variations of it all the time to adapt to the needs of my cycle (use the table I created for you on page 37–38). It always delivers on taste and always makes me feel so nourished and nutritionally topped up because it contains the perfect balance of complete protein, omega-3 fatty acids, complex carbs, fibre and micronutrients for a happy hormonal system.

I love that I can tailor this recipe to what I have in stock and can reinvent it each time by simply changing the vegetables, seasoning and grains. Sometimes I have quinoa if I'm in my follicular or ovulatory phase or switch to brown rice or buckwheat if I'm in my luteal or during my period. I also like to add in some root veggies, like sweet potato or squash, during the colder months of my cycle for extra fibre and complex carbohydrate to suit my increased metabolism.

Serves 2

2 wild salmon fillets
400g tin butter beans, drained, rinsed and dried (or sub any beans)
1 pepper (any colour), roughly chopped
Handful of cherry or vine tomatoes
1 red onion, roughly chopped
2 tsp za'atar (or switch to Italian seasoning)
Pinch each of sea salt and black pepper
Drizzle of extra-virgin olive oil, for cooking
200g quinoa, buckwheat or brown rice, cooked according to the packet instructions
1 lemon slice
Handful of mixed leafy greens and fresh herbs

For the simple salmon marinade

1 garlic clove, grated
1 tbsp raw honey
1 tbsp extra-virgin olive oil
Juice of ½ lemon
Pinch of sea salt and freshly ground black pepper

1. Preheat the oven to 200°C.
2. Combine all the ingredients for the marinade and use to coat the salmon fillets and set to one side.
3. Place the beans, pepper, tomatoes and onion in a roasting tin, sprinkle with your chosen seasoning, plus sea salt and pepper, drizzle with extra-virgin olive oil and roast for 12 minutes. Remove from the oven, place the salmon fillets in the middle and roast for a final 12–15 minutes, or until the fish is cooked through.
4. Serve in bowls with the cooked grains, lemon slice, mixed leafy greens and herbs.

Protein Power Bowl

This classic chopped salad is the definition of a hormone-healthy meal: it's high in complete protein from the organic roast chicken and boiled egg and, therefore, very supportive of stable blood-sugar levels, an essential pillar of your hormone-balancing nourishment guidance and the aim of the game when it comes to supporting all your body systems so they can carry out their job roles efficiently for you. It's also perfectly balanced with healthy fats, fibre and small but mighty micronutrients for a meal that ticks all the hormone health boxes.

Please note this recipe requires precooked chicken, so it's an ideal lunch on a Monday, so that you can put Sunday-roast leftovers to good use (or use my chicken cooking guidance below). But you can also easily swap chicken for tuna, smoked salmon, tofu or legumes, if you prefer.

Serves 2

2 servings of cooked chicken or poached and shredded chicken (see method below)
1 avocado
¼ red onion
⅓ cucumber
Handful of cherry or vine tomatoes
3 tbsp quinoa (or sub buckwheat or any bean), cooked according to the packet instructions
80g feta or goat's cheese
Handful of crushed walnuts
Handful of chopped lettuce or any mixed leafy greens
Handful of basil
2 eggs, hard-boiled
1 tbsp kimchi, optional

For the dressing
4 tbsp extra-virgin olive oil
1 tbsp organic apple cider vinegar
Small handful of finely chopped basil
Pinch each of sea salt and freshly ground black pepper

1. Combine all the ingredients for the salad dressing.
2. Chop the cooked chicken into very small chunks and divide between two bowls.
3. Chop the avocado, red onion and cucumber and divide between the bowls, along with the tomatoes and quinoa.
4. Drizzle with the dressing and top each bowl with crumbled feta or goat's cheese, crushed walnuts, leafy greens, basil, a hard-boiled egg and kimchi (if using).

TIP: Feel free to adapt this salad and make it your own to suit your cycle needs. For example, when you are in your luteal or menstrual phase, you could lighten it by skipping the quinoa or make it slightly more dense by adding roasted sweet potato if you need root-veggie nourishment.

HOW TO POACH AND SHRED CHICKEN

1. Add boneless skinless chicken breasts to a large pot of water and bring to the boil.
2. Boil for 10 minutes until the chicken is no longer pink.
3. Remove from the pot and let cool for a few minutes.

Sticky Salmon Egg–fried Rice

As your metabolism is faster during the second half of your cycle (luteal and menstrual) because your body is using more energy than other phases and, therefore, more nutrients (such as B vitamins and magnesium to prep for your period), your body tends to require slightly more protein, carbs and fats to function optimally. Help your hormones by feeding it this nourishing, high-protein meal made up of protein and omega-3 fatty acids to support hormone production and balance and slow-release complex carbs that will help prevent unhelpful sugar cravings and keep hunger at bay, while also blunting cortisol to help calm your nervous system and promote sleep.

Serves 2

100g brown rice
100g frozen peas
80g cavolo nero, kale or spinach
3 spring onions, chopped
15g fresh coriander, chopped
1 egg
1–2 tbsp tamari (or to taste)
1 tsp garlic granules
Pinch of freshly ground black pepper
1 tbsp toasted sesame oil
⅓ tsp dried chilli flakes (optional but recommended!)
1 tbsp extra-virgin olive oil or cold-pressed sesame oil
2 wild salmon fillets

For the marinade
2 tbsp honey
2½cm piece of fresh ginger
2 garlic cloves
3 tbsp tamari
1 tbsp sesame seeds (black and white)

1. Cook the rice according to the packet instructions.
2. When it is almost cooked and while there is still a little bit of moisture in the pan, add the peas, leafy greens, spring onion and coriander. Cook for a minute before cracking an egg in, mixing it into the rice, along with the tamari.
3. Add the garlic granules, pepper, sesame oil, and chilli flakes and cook for a further few minutes. Set aside with the lid on to keep warm.
4. Combine all the ingredients for the salmon marinade.
5. In the meantime, heat the extra-virgin olive oil in a pan over a medium–high heat. When hot, add the salmon fillets, skin-side down, fry for a minute and then pour the marinade over the top. As it starts to bubble, baste the salmon in the marinade.
6. Cook the fish for 4 minutes on each side, or until cooked through, continuing to cover the salmon in the sauce as you go. (If the sauce caramelises too quickly, add a splash of water to the pan to loosen it.)
7. Remove the salmon from the pan, add to a bowl with the egg-fried rice and drizzle with the remaining sauce in the pan.

Staple Tuna and Egg Blood-sugar-balance Salad

Everyone has a few go-to salads in the bank, and this is definitely one of mine. It never fails to renourish me with two types of complete protein for maximised amino-acid content, healthy fats, phytonutrients and fibre for balanced blood-sugar levels and topped-up nutrient stores. Something I really love about this one is how non-recipe-like it is – it's more a collection of nutrient-dense hormone-supportive ingredients thrown into a bowl. If you are super short on time, you could skip the dressing and go for a simple extra-virgin olive oil and balsamic-vinegar dressing instead.

Serves 2

2 eggs
2 small beetroots
1 avocado
1 spring onion
Handful of cherry or vine tomatoes
400g tin mixed beans, drained and rinsed
2 handfuls of mixed spinach and rocket
100g puy lentils (tinned)
2 heaped tbsp olives
2 x 60g tins tuna
A few crushed walnuts

For the Dijon dressing
1 garlic clove, grated
2 tbsp red wine vinegar
2 tbsp extra-virgin olive oil
2 tsp Dijon mustard
Generous pinch of Italian seasoning
Pinch of freshly ground pepper

1. Place the eggs in a pan of boiling water, bring to the boil and then simmer for 8–10 minutes (or to your liking).
2. Meanwhile, chop the beetroots, avocado, spring onion and tomatoes and pat dry the mixed beans
3. Prepare the Dijon dressing, by combining the ingredients together in a bowl.
4. Divide the greens between two bowls, along with the lentils, beans and all the nourishing plant ingredients.
5. Finish with the tuna and crushed walnuts, drizzle with the Dijon dressing and place the sliced egg on top.

Magnesium-rich Walnut Pesto Spelt Spaghetti

Greens are rich in magnesium and when paired with carbohydrates are incredibly calming on your nervous system. Here, this winning combo blunts cortisol, helping you to wind down in the evening, making this a hormone-hero dinner for your 'food-is-medicine' toolkit.

I'm half Maltese and my Mediterranean blood loves nothing more than a delicious bowl of pasta. It gives me so much joy, and that joy is heightened to new levels when I know it's doing so much good for my health. I recommend giving spelt spaghetti a go if you haven't already; it's kinder on digestion as it's high in gut-supportive fibre and plant-protein, and it's very tasty, too.

This pesto freezes well, so you could double up and freeze a batch for another day when you are short on time and need a tasty, hormone-supportive sauce to stir into pasta. It keeps for four days in a glass, airtight container in the fridge and freezes up to six months.

Serves 2

2 servings of spelt spaghetti
Big handful of spinach
400g tin haricot beans, drained and rinsed

For the walnut pesto

60g basil
30g flat-leaf parsley
60g walnuts
60g Parmesan
2 garlic cloves
125ml extra-virgin olive oil
Juice of ½ lemon
Pinch each of sea salt and freshly ground black pepper
¼ tsp dried chilli flakes

1. To make the pesto, place all the ingredients in a food processor and blitz.
2. Cook the spaghetti according to the packet instructions, drain and stir in the spinach, and haricot beans (or sub butter beans for your choice of protein, or prawns or chicken), along with the pesto, while the pasta is still hot.

TIPS: If you fear carbs, remember, insufficient starch intake can suppress luteinising hormone and ovulation. Your hormones quite literally *need* carbs to be balanced, so don't fear them – eat them in moderation as part of your healthy, diverse diet.

This recipe works well served with chicken or prawns to give it a complete protein boost.

Speedy Three-ingredient Greek-yoghurt Flatbreads

Say goodbye to shop-bought flatbreads with lists of ingredients you can't even pronounce. These super-simple-to-make versions use only a few simple ingredients, they're rich in protein (thanks to the Greek yoghurt) and make the ideal base for lunch or dinner.

The best thing about these flatbreads is their versatility. I love adding chilli jam for a subtle kick, za'atar for Moroccan-style sharing food or mango chutney to go with my chicken tikka thighs (see page 124) or dunking them into curries, which also makes them adaptable for all phases of your cycle.

Just add your choice of protein from chicken to salmon, grilled halloumi or tofu, pickled onions, or kimchi, a handful of bitter herbs to aid digestion and some hummus, then tuck right in.

Makes 4 small flatbreads

150g self-raising flour (or add 1½ tsp to plain/gluten-free flour), plus extra for dusting
150g Greek yoghurt
Pinch of sea salt
1 heaped tsp chilli jam or mango chutney (optional)
Drizzle of extra-virgin olive oil, butter or a little ghee, for cooking

1. Add the flour, Greek yoghurt and sea salt to a big bowl and combine. Add the chilli jam or mango chutney at this stage, too, if using.
2. Combine the ingredients and roll into one big ball, then leave to rest for 15 minutes if you have time (if you don't, it still works).
3. Sprinkle some flour on a board or a clean worktop and divide the ball into four small equal-sized balls. Using a rolling pin, roll each ball out into a flatbread shape.
4. Heat some oil, butter or ghee in a pan over a medium–high heat and cook the flatbreads one at a time for around 20–40 seconds until golden brown, before flipping (don't cook them for too long as they easily burn).
5. Load with veggies and protein or dip into your fave dish and enjoy.

Quick Pickled Onions

To make yourself a batch of quick pickled onions to amplify your meals, finely chop a large onion, add to a glass airtight container and set to one side.

In a small saucepan, add 100ml water, 100ml apple cider vinegar, 2 tablespoons maple syrup or honey, 1 teaspoon sea salt, 1 teaspoon pepper and a pinch of dried chilli flakes. Bring the mixture to a gentle simmer over medium heat, then carefully pour the mixture into the jar over the onions. Allow to cool and then store in the fridge for up to 14 days.

Curried Mango, Prawns and Grains

Protein, complex slow-burning carbs, cruciferous vegetables and juicy antioxidants – this bowl contains all the nourishing key components to make up a hormone-balancing meal to keep unwanted hormones and toxins out and to promote hormonal harmony. Its particularly supportive energy production is due to a range of B vitamins (such as B12 and folate), which help to replenish red blood cells. Prawns are also a useful source of the trace minerals iodine, zinc and selenium. We need iodine to support the correct function of the thyroid gland, while zinc and selenium support the immune system.

If you can make this recipe with black rice, so much the better because it contains high amounts of phytochemicals, particularly antioxidant anthocyanins.[119] In fact, it contains more antioxidants than any other rice!

Serves 2

2 servings of black or brown rice (approximately 250g)
150g broccoli
Drizzle of extra-virgin olive oil
2 garlic cloves, finely chopped
1 tbsp medium curry powder
2 tbsp mango chutney
150g uncooked prawns
2 handfuls of spinach
1 avocado, sliced
A few small radishes, thinly sliced
Handful of mango chunks
2 spring onions, finely chopped
Small handful of fresh coriander

To finish

Drizzle of cold-pressed sesame oil
Sprinkle of sesame seeds
Roughly chopped cashews

1. Cook the rice according to the packet instructions.
2. Place the broccoli in a steamer basket over a pot with 2.5cm water, bring the water to a simmer and steam for approximately 4 minutes. Be careful not to oversteam it, as you want to retain as many nutrients as possible.
3. Meanwhile, heat the extra-virgin olive oil in a pan over a medium heat, add the garlic and curry powder and cook for a few minutes before stirring in the mango chutney. Cook for a few more minutes.
4. Stir the prawns into the mix and heat them through, until cooked.
5. Divide the spinach between two bowls, followed by the rice and serve the prawns on top, straight from the pan.
6. Finally, add the avocado slices, the steamed broccoli, radish slices and mango chunks and top with the spring onions, coriander, chopped cashews, sesame seeds and a drizzle of cold-pressed sesame oil, if you fancy going all out for the extra nutrients!

Tamari Tuna, Tomatoes and Tenderstem Broccoli

This is another one of those simple, healthy dishes that uses wholefoods and store-cupboard staples that taste as good as they make you feel, and a prime example of a hormone-balancing dinner and a tasty weeknight staple. Plus, my husband James loves it, which is always a winner!

I especially love making this straight after my period to help replenish nutrient stores, as tuna is a lean complete protein and a good source of iron, and paired with the iron-rich lentils and high fibre content, it makes this ideal to whip up during the week, when speedy nourishment is my main goal.

Serves 2

Handful of cherry or vine tomatoes
1 tbsp extra-virgin olive oil
2 tuna steaks
200g tenderstem broccoli
220g tin lentils, drained and rinsed
Handful of fresh coriander (approximately 30g)
2 spring onions, finely chopped
2 tbsp cashews, finely chopped or crushed

For the tuna marinade

2 tbsp tamari
2 garlic cloves, grated
2½cm piece of fresh ginger, grated
1½ tbsp raw honey
Pinch of dried chilli flakes

1. Preheat the oven to 180°C.
2. Place the tomatoes on a lined baking tray with a tiny drizzle of the extra-virgin olive oil and place in the oven for 15 minutes.
3. Combine the ingredients for the tuna marinade.
4. Heat the remaining oil in a pan over a medium heat. Add the tuna steaks to the pan, pour over the marinade and pan-fry for a few minutes on both sides (depending on how thick they are). Tuna should be slightly pink in the middle and cooked on the outside – but cook for longer if you prefer.
5. As soon as you put the tuna steaks in the pan, place the broccoli in a steamer basket over a pot with 2.5cm water, bring the water to a simmer and steam for approximately 4 minutes. Be careful not to oversteam it, as you want to retain as many nutrients as possible.
6. Heat the lentils, remove the tomatoes from the oven and add to a bowl with the tuna steaks, tenderstem and coriander. Top with the spring onions and crushed cashews and enjoy.

Miso Salmon and Mushroom Traybake

Omega-3 fatty acids in your hormone-balancing diet are crucial because they provide essential building blocks for your brain, cells and hormones. These cannot be produced by the body, so we must get them from food or supplements.

I talk to clients daily about the importance of increasing omega 3s from food to help naturally prevent and reduce PMS (and downregulate inflammation associated with acne breakouts), and I've seen impressive results both in my practice and in my own cycle. In fact, results of a meta-analysis showed that compared to a control panel the daily consumption of omega 3 reduced the severity of symptoms, including depression, nervousness, anxiety and lack of concentration and may also reduce the somatic symptoms of PMS, including bloating, headache and breast tenderness, and results continued to improve the longer omega 3 was consumed.[120]

The miso in this recipe is also a key hormone-balancing ingredient that some people are scared to eat because of outdated research based on genetically modified soy. But from my perspective, it's a food-medicine ingredient I use therapeutically to help my clients reduce hormone-imbalance symptoms during perimenopause and menopause. Science backs this up, concluding that isoflavones found in quality organic, fermented (non-GMO) soy products effectively reduce the frequency of hot flushes and fatigue, enhancing quality of life during this time.[121]

However, the effectiveness is determined by how well your gut can absorb them, which is why optimising gut health is always key to any nutritional enhancements you make.

Serves 2

400g tin chickpeas, rinsed and drained
200g mushrooms, sliced
1 courgette, chopped
1 tbsp toasted sesame oil
Pinch of freshly ground black pepper
2 wild salmon fillets
Big handful of cavolo nero
Small handful of fresh coriander (approximately 15g)
Pinch of dried chilli flakes (optional)

For the miso marinade
2 tsp white miso paste
5cm piece of fresh ginger, grated
2 tbsp tamari
1½ tsp toasted sesame oil

1. Preheat the oven to 190°C.
2. Combine all the ingredients for the miso marinade.
3. Pat dry the chickpeas and place in a roasting tin with the mushrooms and courgette.
4. Drizzle the toasted sesame oil over the vegetables, season with pepper and toss together. Place in the oven for 15 minutes.
5. While the vegetables are roasting, prepare the salmon by coating with the miso marinade.
6. Remove the roasting tin from the oven and place the salmon fillets in the middle. Roast uncovered for roughly 12–16 minutes, or until visibly cooked.
7. Around 5 minutes before you are ready to serve, add the cavolo nero and top with the coriander and chilli flakes (if using).

Honey-mustard Chicken Thighs with Rosemary New Potatoes

Quality lean protein from meat such as chicken includes essential amino acids that produce the neurotransmitters dopamine and serotonin, which are linked to supporting the nervous system and improving cognitive health.[122] This one-tray chicken recipe is balanced with new potatoes for carbohydrates, which helps to blunt cortisol to aid sleep, while tenderstem broccoli for gut-health fibre makes it an ideal neurotransmitter-supporting balanced meal to eat in the evening for your brain and beyond.

Serves 2 (plus a little one)

300g new potatoes
1 tbsp extra-virgin olive oil
2 tsp dried rosemary
Pinch each of sea salt and freshly ground black pepper
4–6 organic chicken thighs
100g cherry tomatoes
200g tenderstem broccoli
1 tbsp honey
2 spring onions, finely chopped
Spinach and rocket, to serve (optional)

For the honey-mustard marinade

2 tbsp raw honey
1 tbsp Dijon mustard
1 tsp wholegrain mustard
2 garlic cloves, grated
3 tbsp extra-virgin olive oil
1 tsp dried rosemary
1 tsp grated lemon zest, plus a big squeeze of lemon
Pinch each of sea salt and freshly ground black pepper

1. Preheat the oven to 200°C.
2. Slice the new potatoes into halves, then quarters (the smaller the chunks, the quicker they will cook) and transfer them to a roasting tin. Drizzle with the extra-virgin olive oil, add the rosemary and the sea salt and pepper and pop in the oven for 25 minutes.
3. Meanwhile, combine all the ingredients for the marinade and pour this over the chicken thighs in a bowl. Toss together.
4. Add the honey-mustard thighs, tomatoes and broccoli to the roasting tin (with the potatoes), drizzle with the honey and pop in the oven for a further 30–35 minutes, until the chicken is golden and slightly browned on the edges, and the juices run clear when pierced with a knife.
5. Serve with the spring onions on top, along with spinach and rocket, if desired.

TIP: Leave your potatoes to cool after cooking as this releases the resistant starch to lower glycaemic response and feed the gut microbiome.

My Mama's Slow-cooked Jerk-style Chicken, Rice and Beans

This one-pot dish is rich in soluble fibre (from the beans which your gut microbes feed off and thrive on), complex carbs (from the wholegrain rice to blunt cortisol and support HPA axis, so helping to manage anxiety and stress) and a complete amino-acid rich protein (from the chicken, to support tryptophan production to improve your sleep).[123] It is incredibly easy to make, spicy and creamy in equal measure and bursting with flavour. I can't take the credit for this one, though. It's my mum's speciality and really tastes like home for me. I hope you find it as delicious and wholesome as I do.

This hearty, slow-cooked meal is ideal just before or during your period in the autumn and winter of your cycle, as it will replenish you with the ingredients your body needs as your hormones plummet and you need all the warming nourishment you can get into your system.

You'll need a slow cooker to make this.

Serves 6

6 organic, skinless, boneless chicken thighs (approximately 500–600g)
2½ tbsp jerk seasoning (if you like light–medium spice)
200g frozen peas
1 tbsp extra-virgin olive oil or ghee
2 red onions, chopped
5 garlic cloves, halved
400ml full-fat coconut milk
400g tin kidney beans, drained and rinsed
400g tin black beans, drained and rinsed
400g wholegrain rice
600ml chicken stock
60g fresh coriander
1 tbsp Greek yoghurt, to serve

1. Marinate the chicken thighs by coating them with the jerk seasoning in a bowl. Leave overnight if you can, or for as long as possible before cooking.
2. Pour the frozen peas into a bowl, cover with boiling water and set aside to thaw.
3. Heat the extra-virgin olive oil or ghee in a pan over a medium heat and place the jerk-chicken thighs in the pan, flattening them down.
4. Add the red onions and garlic halves to the pan and keep turning the thighs to lightly brown them off.
5. Transfer everything from the pan to a slow cooker and pour in the coconut milk, along with the beans.
6. Add extra jerk marinade if you want it extra spicy! Now give everything a good stir. Set the slow cooker to low for a minimum of 4 hours (I usually cook it for 6 on low).
7. Around 2 hours before you want to eat, add the uncooked rice and the stock.
8. An hour before, drain the water from the bowl of peas and stir them into the pot.
9. Just before serving, add the roughly chopped coriander and a dollop of Greek yoghurt!

I recommend buying pre-made jerk seasoning if you can for speed and ease, but if you want to make your own, these are the herbs and spices you'll need: 1 tsp dried thyme, 1 tsp ground cumin, 2 tsp cayenne pepper, 2 tsp garlic granules, 1 tbsp smoked paprika, 1 tsp allspice, 1 tsp salt, 1 tsp pepper, ½ tsp dried chilli flakes, ½ tsp nutmeg, ½ tsp ground cinnamon. Add extra-virgin olive oil to form a paste ready to marinate the chicken thighs.

Thai Green Salmon Noodles

My Thai green salmon noodles are seriously tasty, easy to make and you'll only need twenty minutes from start to finish. In fact, if you're short on time, you can substitute prepared stir-fry vegetables for the fresh vegetables listed below.

This recipe is rich in the nutrients needed to support your nervous system, while keeping your hormones happy, too. There are omega-3 fatty acids, amino acids and protein to support your brain (from the wild salmon), phytoestrogens (from the edamame beans) for sex-hormone balance, wholegrain fibre (from the noodles) and a big boost of anti-inflammatories (from the lemongrass and fresh coriander).

It's the sort of nourishing weeknight dinner you'll keep coming back to and makes a great recipe for when hosting friends, too, because it looks and tastes like it involves far more ingredients and effort than it actually does!

Serves 2

2 wholegrain rice-noodle nests or 2 servings of soba noodles
1 tbsp olive or sesame oil
1 yellow pepper, chopped
1 large red onion, chopped
150g tenderstem broccoli, chopped
400ml full-fat coconut milk
5 medium-sized tomatoes, chopped
1 lemongrass stalk
2 tbsp fish sauce
½ vegetable stock cube
3 tbsp Thai green curry paste
150g frozen organic edamame beans
Juice of 1 lime
30g fresh coriander, plus extra to garnish
2 wild salmon fillets
Big handful of fresh spinach

1. Place the wholegrain rice noodles in a bowl, cover with boiling water and set aside (or prepare the soba noodles according to the packet instructions).
2. Heat the oil in a pan over a medium heat. Once hot, add the chopped vegetables and stir-fry until soft.
3. Pour in the coconut milk, along with the tomatoes. Grate half the lemongrass stalk (to prepare it, just chop the hard end off and remove the outer layers) and finely chop the rest and add to the pan with the fish sauce. Stir well
4. Crumble in half a vegetable stock cube and add the Thai green curry paste. Stir well again.
5. Add the edamame beans, the lime juice and coriander (with stalks).
6. Place the salmon fillets in the middle, skin-side up and pop a lid on the pan. Leave to simmer for 6 minutes.
7. Using a spoon, very gently pull and peel back the salmon skin to remove it, then chop the fillets into small chunks.
8. Drain the noodles and add to the pan, along with a handful of spinach. Cover again for a final 2 minutes before serving with a garnish of coriander.

SPOTLIGHT:
Phytoestrogens found in organic edamame are helpful to add to your diet during perimenopause and menopause to relieve symptoms caused by fluctuating oestrogen.[124] Plus, they are a great complete plant protein.

Creamy Mushroom and Pea Pasta

This recipe gives all the moreish creaminess you'd usually find in a cheese sauce but minus the dairy, as I've used coconut yoghurt. Super speedy to throw together and delivers a big hit of nutrients, including protein from the peas (one of the most underrated and accessible vegetables) and vitamin D from the mushrooms.[125] Vitamin D (which is, in fact, a hormone) is essential for maintaining a balanced hormonal system. It protects us from running out of serotonin, and helps to regulate the production of adrenaline, noradrenaline and dopamine in the brain. Getting vitamin D from sunlight is important, but also challenging, depending on the time of year, the country you live in and how much time you actually spend outside absorbing it from sunlight. So ensuring your diet includes vitamin D sources is key; mushrooms are one, as well as salmon, egg yolk and grass-fed red meat. I would also recommend working with a naturopathic nutritionist to discuss whether supplementation is necessary for your hormone health and goals.

Serves 2

200g dried spelt pasta
1 tbsp extra-virgin olive oil
Pinch each of sea salt and freshly ground black pepper, plus extra salt for cooking the pasta and extra pepper for serving
Pinch of dried chilli flakes (optional)
100g frozen peas
200g chestnut mushrooms, halved
3 garlic cloves, grated
200g spinach
3 heaped tbsp coconut yoghurt (or Greek yoghurt can be used)
Handful of flat-leaf parsley, chopped

1. Cook your pasta according to the packet instructions, adding a drizzle of the extra-virgin olive oil and sea salt to the water
2. Pour boiling water over the frozen peas and set aside to thaw.
3. While the pasta is cooking, sauté the mushrooms and garlic in the remaining oil, over a medium heat with the salt, pepper and chilli flakes.
4. When the mushrooms are soft, add the spinach and defrosted peas and cook for a further few minutes.
5. Drain the pasta and add to the mushroom/spinach pan, before stirring in the yoghurt.
6. Serve with a big sprinkle of parsley and a big twist of black pepper.

SPOTLIGHT:
Remember to leave mushrooms out in the sun for at least 15 minutes before cooking to increase the vitamin D content.

TIP: This recipe works well served with chicken or prawns to give it a complete protein boost.

Adrenal Support Tonic

When you're stressed, your adrenal glands use more vitamin C than they would on a regular basis. Beyond regulating cortisol production, vitamin C's antioxidant properties are also needed to fuel the adrenal glands, protect against oxidative stress and boost the immune system.

So why not swap your afternoon caffeine habit for this deeply hydrating, cortisol-balancing, magical, vitamin-and-mineral-rich elixir? Containing electrolytes (including calcium, magnesium and potassium), it efficiently replenishes lost fluids and salts and the vitamin C that is zapped in times of stress. This delicious tonic is functional food at its finest and helps naturally treat stress and fatigue, while regulating blood pressure from within by renourishing your depleted adrenals whenever they are in need.

Coconut water is a diuretic and helps to regulate blood sugar, making it ideal for PCOS sufferers or anyone on their hormone-healing journey who wants to take hydration to the next level.

Serves 1

200ml organic raw coconut water
1 serving (5g) of unflavoured collagen powder
¼ tsp quality sea salt or Celtic salt
Juice of ½ orange or lemon

1. Pour the coconut milk into a glass and stir in the collagen powder, salt and orange or lemon juice.
2. Stir well before drinking. Alternatively, add to a blender and blitz with ice before serving.

8. Putting it into Practice

A QUICK GUIDE TO NUTRIENT BASICS FOR HORMONE BALANCE

We've looked at the nutrient basics in various parts of the book already, but here's a quick at-a-glance guide for you to refer to when you are meal planning to help you take the nutrient profile of your meals to the next level.

PHYTOCHEMICALS

Phytochemicals are natural chemicals in plants, and many have antioxidant, anti-inflammatory and antimicrobial properties that benefit your hormonal health. Vegetables are packed with phytochemicals and their high antioxidant value means that they play a protective role in reducing levels of inflammatory markers, which is key for preventing and treating hormonal-imbalance-related conditions such as PCOS, endometriosis and PMDD.

DIETARY FIBRE

Fibre is the foundation to good gut health – and good gut health is the foundation to good hormonal health, making fibre absolutely essential when it comes to hormone balance. Think legumes, brown rice and sweet potatoes.

FATTY ACIDS

Fat is essential for hormone production and balance. Fats are made up of fatty acids, and there are two main types: saturated and unsaturated. Monounsaturated are considered the healthiest and polyunsaturated (aka essential fatty acids) are important because our bodies can't make them, meaning we must get them from our diets, from oily fish such as salmon, mackerel, anchovies, sardines and herring (or, when it comes to plant-based sources, flaxseed, chia seeds and walnuts are rich sources).

OPTIMISE YOUR OMEGA 3 TO 6 RATIO

Omega 3 and omega 6 are polyunsaturated fats. Scientists suspect that a distorted ratio of these fatty acids may be one of the most damaging aspects of the Western diet, and that a diet high in omega 6s but low in omega 3s increases inflammation, while one that includes higher amounts of omega 3 reduces inflammation. The best way to achieve a health-promoting ratio of these fatty acids is to consume a diverse, plant-rich diet, quality extra-virgin olive oil, oily fish and nuts and seeds (such as walnuts, chia and flax) and consciously limit UPFs.

But if I were to give you one piece of advice above all, the single most important thing you can do here is to avoid highly refined, processed seed and vegetable oils. These are high in omega 6 and oxidise easily (which means they are less likely to contain free radicals), and therefore promote inflammation. However, the caveat, cold-pressed and organic, quality seed oils are not 'bad' and can have health benefits when consumed in small amounts, but we require less of them in our diets and more omega 3 from extra-virgin olive oil, for example.

Also be aware of foods you buy regularly that contain these oils, as they tend to be snuck into lots of pre-packaged foods (even marketed 'health' foods).

Omega 3s are important for brain, bone and heart health and play a critical role in your hormonal health. They are hormonal precursors for producing the key sex hormones needed for hormone balance (oestrogen, progesterone and testosterone); and they regulate and lower inflammation, protecting the body's ability to make hormones. Omega-3 fatty acids must come from the diet or by taking a supplement.

A FEW KEY VITAMINS, MINERALS AND MICRONUTRIENTS FOR HORMONE BALANCE

B vitamins, particularly B2, B6 and B12 – because research shows that this trio of vitamins helps to balance oestrogen and progesterone, which can reduce PMS symptoms. Deficiencies in these specific B vitamins can negatively impact reproductive health.[126] Food sources include meat (especially liver), seafood, poultry, eggs, dairy products, legumes, leafy greens, seeds and nutritional yeast.

Zinc regulates your HPA axis aka stress response. It also supports your ovarian follicles to promote healthy ovulation and progesterone production and it's a key player when it comes to the synthesis, transport and action of all your hormones, plus it's essential to heal acne-prone skin. In the case of PCOS, it's important, as it blocks androgens, which are elevated in the condition. The richest sources of zinc include meat, fish and seafood.

Vitamin D should be called a hormone because that's exactly what it is! It's a steroid hormone that regulates more than 200 genes and is essential for healthy insulin sensitivity and for healthy, regular ovulation. It's produced in your skin from sunlight and helps to absorb calcium from your gut microbiome into the bloodstream.

Iodine is essential for thyroid health, but it's also important for treating oestrogen-excess symptoms, including breast pain, ovarian cysts and PMS and PMDD, as it promotes healthy detoxification of oestrogen. Food sources include seaweed (kombu kelp, wakame and nori), eggs and fish such as tuna, cod and prawns.

Magnesium soothes your nervous system and is a powerful stress reliever. It regulates the HPA axis, promotes sleep, plus it activates your GABA receptors, which calms you down. It's a miracle mineral for your periods, as it supports insulin and thyroid hormone and the healthy metabolism of oestrogen, which is why it's so essential when it comes to overall hormone balance.[127, 128] Food sources include leafy greens, avocados, pumpkin seeds, nuts (cashews, almonds, Brazil) wholegrains (brown rice, quinoa, buckwheat), legumes and dark chocolate.

Your Unique Protein Needs For Hormone Balance

Protein is essential for hormone balance. Without adequate protein you cannot keep your hormones in check or give your body what it needs to be healthy. Never underestimate the importance of protein on your plate if you want to balance your hormones – and keep them balanced.

Protein is made of amino acids, the building blocks for female hormones that maintain reproductive health and regulate our cycles, mood, neurotransmitters for mental health and fertility. Protein also keeps your blood-sugar levels steady, which means you will feel more balanced physically and mentally.

Not only does protein provide essential amino acids that your body can't make on its own, your endocrine glands also need it to produce protein-derived hormones, also known as peptide hormones. Peptide hormones play a crucial role in regulating many physiological processes, such as growth, energy metabolism, appetite, stress and reproduction.

Eating enough complete protein ('complete' meaning it has a full amino-acid profile for optimal body functioning) is therefore very important. Meat, fish and eggs are not only wonderful complete-protein sources, but also contain an abundance of essential vitamins, minerals and omega-3 fatty acids to amplify your health.

How do I build a hormone-balanced plate?

As you know by now, it's the combination and diversity of vegetable varieties, complete protein, complex carbohydrates and healthy fats that creates a plate of medicine food that will quite literally feed your body systems and, therefore, nourish your hormones into balance, one meal at a time.

I prefer not to focus on numbers, as it gets in the way of my intuitive-eating method, so instead, I go by visual reference, using my plate or hand guides below.

Take a look at the nutrition cheat sheet I created for you for ideas on what you can add to your plate (see page 218), remembering that each plant and ingredient and plant source contains a different array of vitamins and minerals to support your body with.

Devote a quarter of your plate to complete animal protein or a third to plant-based protein sources (learn about plant protein-combining opposite). Aiming for this amount of protein on your plate at each meal will help you to achieve balanced blood-sugar levels throughout the day, prevent insulin spikes and unhelpful sugar cravings and annoying energy dips, as well as supporting muscle recovery post exercise.

Next, ensure half of your plate is filled with non-starchy vegetables. This is where you get the opportunity to supply your gut with the fibre it needs to thrive, plus an array of

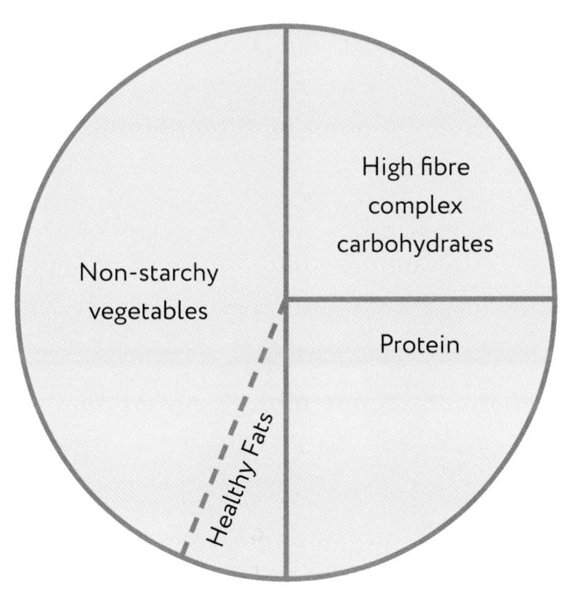

Balanced plate guide

micronutrients to top your nutrient stores up and support your body systems with the anti-inflammatory phytochemicals and plant compounds it requires to be healthy.

Make a quarter of your plate a complex-carbohydrate source. This will also support stable blood-sugar levels, as they're digested more slowly and therefore supply a slower release of glucose into the bloodstream, keeping you feeling satisfied for longer. Remember also that insufficient carbohydrate intake can suppress luteinising hormone and ovulation and can hinder your menstrual cycle.

Finally, add a fist size quantity of healthy fats because they are the essential building blocks for your hormones, helping with the absorption of fat-soluble vitamins, and are integral for brain and skin health, while keeping inflammation at bay.

GO WILD

Opting for organic poultry, grass-fed meat, eggs and wild fish means you protect yourself from consuming the antibiotics and growth hormones given to livestock. Farmed salmon are not only given antibiotics but are often dyed pink to make them more desirable, so opt for wild, always.

Protein Combining

It can be tricky to meet your protein requirements when following a plant-based regime, so try to be mindful of adding a few sources to each meal and enjoy the creativity of including different, delicious, plant-based ingredients to your plate.

Combining plant-based protein sources[129] well will help you to achieve an optimum amino-acid profile, which will make you feel more satisfied from your meal, preventing the feeling of weakness/low energy that often leads to unhelpful sugar cravings. This will also increase the diversity in your diet and, therefore, the range of nutrients you are supporting your system with. More fibre means a happy gut microbiome and more nutrients equals happier hormones. This can be achieved by mixing any two from these three groups:

- Wholegrains
- Nuts/seeds/nut butters
- Legumes

Examples of good combinations include:

- hummus on rye bread
- peanut butter on oatcakes
- vegetable stir-fry with black-eyed beans and gluten-free brown rice noodles

NUTRITION CHEAT SHEET AND SHOPPING LIST IDEAS

COMPLETE PLANT PROTEIN (containing all nine essential amino acids)	• Quinoa • Buckwheat • Organic edamame beans	• Peas • Organic tofu • Organic tempeh • Hemp seeds	• Spirulina • Vegan collagen • Amaranth • Chia seeds
ANIMAL PROTEIN (complete protein)	• Chicken • Turkey • Salmon • Cod • Mackerel • Sea bass	• Herring • Seafood • Organic eggs • Greek yoghurt • Halloumi • Feta	• Cottage cheese • Goats cheese • Bovine collagen • Tuna
NON-STARCHY VEGETABLES (low in carbohydrates)	• Peppers • Mushrooms • Onions • Garlic • Leeks • Spring onions • Beetroots • Green beans	• Carrots • Cabbage • Broccoli and sprouts • Asparagus • Aubergines • Courgettes • Cucumber • Broccoli	• Sea vegetables (seaweed, nori, kelp, dulse) • Cavolo Nero • Artichokes • Spinach • Celery • Tomatoes
STARCHY VEGETABLES/ GRAINS AND LEGUMES (higher in carbs)	• All legumes/lentils • All rice • Quinoa • Buckwheat	• All potatoes, including sweet • Squash • Green bananas	• Barley • Parsnips
HEALTHY FATS	• Extra-virgin olive oil • Avocado oil • Flaxseed oil • Walnut oil	• Avocados • Olives • All nuts • All seeds	• MCT* • Grass-fed butter • Oily fish • Coconut oil
LEGUMES/LENTILS (plant proteins)	• Mixed beans • Chickpeas • Kidney beans • Butter beans • Haricot beans	• Borlotti beans • Red lentils • Puy lentils • Split peas • Pinto beans	• Split peas • Cannellini beans • White beans • Mung beans
COMPLEX CARBS (naturally gluten-free)	• Quinoa • Buckwheat • Millet • Brown rice • Black rice	• Wild rice • Polenta • Sweet potatoes • Squash • Red-lentil pasta	• Chickpea pasta • Pea pasta • Legumes

OMEGA-3 SOURCES	• Salmon • Mackerel • Anchovies	• Sardines • Herring • Walnuts	• Chia seeds • Flaxseed • Seaweed and algae
ANTIOXIDANT SOURCES	• All berries • Sweet potatoes • Purple sweet potatoes • Purple sprouting broccoli	• Peppers • Red cabbage • All dark, leafy greens • Red onions • Dark chocolate	• Black rice • Beetroot • All herbs and spices • Artichokes • Extra-virgin olive oil
CRUCIFEROUS VEGETABLES	• Broccoli • Kale • Cabbage • Brussels sprouts	• Pak choy • Cauliflower • Turnips • Radishes	• Rocket • Broccoli sprouts • Watercress • Cavolo nero
NUTS AND SEEDS	• Almonds • Walnuts • Brazil • Hazelnuts • Macadamia	• Pecans • Pine • Chia • Flax • Hemp	• Sunflower • Pumpkin • Tahini • Nut butters
HERBS AND SPICES	• Ginger • Turmeric • Garlic • Cinnamon	• Coriander • Rosemary • Parsley • Mint	• Dill • Cayenne pepper • Sage • Fenugreek
PROBIOTICS	• Yoghurt – natural/ Greek • Kefir	• Sauerkraut • Tempeh • Kimchi	• Miso • Kombucha • Natto
PREBIOTICS	• Chicory root • Dandelion greens • Jerusalem artichokes • Garlic	• Onions • Leeks • Asparagus • Bananas • Barley	• Oats • Apples • Cocoa • Flaxseed • Seaweed

* (medium-chain triglyceride) oil

FOURTEEN-DAY FOLLICULAR-PHASE MEAL PLAN (DAYS 1–14)

From the first day of your period to day 15, when the luteal phase begins.

	DAY 1	DAY 2	DAY 3	DAY 4
BREAKFAST	Porridge with protein powder berries and hemp seeds	Boiled egg slices, mashed avocado on a rosemary seeded cracker	Greek/live yoghurt, Nutty, Seedy, Slightly Spiced Granola	Greek/Live yoghurt, Nutty, Seedy, Slightly Spiced Granola
LUNCH	Healing 'Hug-in-a-Bowl' Chicken Soup	Restorative Beetroot Soup	Multivitamin Gut-nourish Bowl	Glowing Green Miso Noodle Soup
DINNER	Replenishing Classic Chili	Roasted Cauliflower, Cashew and Buckwheat Bowl with Tahini	Supercharged Sweet Potato and Smoky Beans	Creamy Coconut Chickpea Curry

	DAY 8	DAY 9	DAY 10	DAY 11
BREAKFAST	Greek/Live yoghurt, Nutty, Seedy, Slightly Spiced Granola	Boiled egg slices, mashed avocado on a rosemary seeded cracker (sprinkled with sprouts optional)	Greek/live yoghurt, Nutty, Seedy, Slightly Spiced Granola	Creamy Greens, Happy-hormone Smoothie
LUNCH	Orangey Kale and Feta Lentils	Staple Tuna and Egg Blood-sugar-balance Salad	Ultimate Detox Reset Salad	Beetroot Beauty-food Bowl
DINNER	Curried Mango, Prawns and Grains	Miso Mushroom Salmon Traybake	Tamari Tuna, Tomatoes and Tenderstem Broccoli	Quinoa-crusted Aubergine Parmigiana

DAY 5	DAY 6	DAY 7
BREAKFAST Turmeric Messy Eggs and Avocado	BREAKFAST Greek/live yoghurt, Nutty, Seedy, Slightly Spiced Granola	BREAKFAST The Balance Pot
LUNCH Avocado, Herb and Walnut Skin-glow Sourdough	LUNCH Sweet Potato and Chickpea 'Happy-gut' Wraps	LUNCH Speedy Italian Butter-bean Stew
DINNER My Mama's Slow-cooked Jerk Chicken, Rice and Beans	DINNER Cajun Maple Chicken, Butter Beans and Greens	DINNER Thai Green Salmon Noodles

DAY 12	DAY 13	DAY 14
BREAKFAST Kefir Blueberry Breakfast Smoothie	BREAKFAST Greek/live yoghurt, Nutty, Seedy, Slightly Spiced Granola	BREAKFAST The Balance Pot
LUNCH Immune-boosting Antioxidant Salad	LUNCH Favourite Fig Greek Salad	LUNCH Liver-detox Shredded Asian Satay Salad
DINNER 'Sunshine-in-a bowl' Salmon Salsa	DINNER Rainbow Summer Tofu and Quinoa Spring Rolls	DINNER Liver-loving Ginger and Lemongrass Sea Bass

Seed cycling: add 1 tablespoon ground flaxseed and 1 tablespoon pumpkin seeds to breakfast, lunch or dinner.

Drink a cup of my immunity tea every morning (see page 170) and 2 litres fresh filtered water daily (herbal teas count).

Make your porridge with a combination of gluten-free oats and buckwheat to increase its nutrient profile.

FOURTEEN-DAY LUTEAL-PHASE MEAL PLAN (DAYS 15–28)

After ovulation until day 1 of period.

DAY 15	DAY 16	DAY 17	DAY 18
BREAKFAST Kefir Blueberry Breakfast Smoothie	BREAKFAST Creamy Greens Happy-hormone Smoothie	BREAKFAST Avocado, Herb and Walnut Skin-glow Sourdough	BREAKFAST Greek/live yoghurt, Nutty, Seedy, Slightly Spiced Granola
LUNCH Speedy Three-ingredient Greek-yoghurt Flatbreads with chicken, hummus and salad	LUNCH Every-season Tomato, Fennel and Buckwheat Soup	LUNCH Protein Power Bowl	LUNCH Honeyed Halloumi and Borlotti Bean Salad
DINNER Superfood Buckwheat and Thyme Skin-soothing Salad	DINNER My Best-ever Dhal	DINNER Harissa, Butter Bean and Red Pepper Orzo	DINNER Go-to Garlic-rubbed Roast Chicken with sweet potatoes and roasted seasonal vegetables

DAY 22	DAY 23	DAY 24	DAY 25
BREAKFAST Greek/live yoghurt, Nutty, Seedy, Slightly Spiced Granola	BREAKFAST The Balance Pot	BREAKFAST Turmeric Messy Eggs with mashed avocado on a rosemary seeded cracker	BREAKFAST Greek/live yoghurt, Nutty, Seedy, Slightly Spiced Granola
LUNCH Creamy Cauliflower, Leek and Cashew Soup	LUNCH Seasonal Prebiotic Traybake with Tahini Dressing	LUNCH 'PMS-free' Winter Squash Salad	LUNCH Restorative Beetroot Soup
DINNER Magnesium-rich Walnut Pesto Spelt Spaghetti	DINNER Peanut and Ginger Noodle Stir-fry	DINNER Sticky Salmon Egg Fried Rice	DINNER 'Bring-you-back-to-life' Comforting Cauliflower Curry

DAY 19	DAY 20	DAY 21
BREAKFAST Eggs, avocado, with rosemary seeded crackers	BREAKFAST Turmeric Messy Eggs with mashed avocado on a rosemary seeded cracker	BREAKFAST Porridge with protein powder berries and seeds
LUNCH Creamy Coconut Chickpea Curry	LUNCH Love-you-back Greens and Beans Soup	LUNCH Sweet Potato Salad with Medicinal Ginger Dressing
DINNER 'Make-it-your own' Simple Salmon Traybake, served with brown rice	DINNER 'Friday Night' Chicken Tikka 'Fakeaway'	DINNER Harissa Cod, Beets and Bean Skin-support Traybake

DAY 26	DAY 27	DAY 28
BREAKFAST Turmeric Messy Eggs with mashed avocado on a rosemary seeded cracker	BREAKFAST Greek/live yoghurt, Nutty, Seedy, Slightly Spiced Granola	BREAKFAST The Balance Pot
LUNCH Herby Halloumi and Mediterranean veggies served in a wrap	LUNCH Multivitamin Gut-nourish Bowl	LUNCH Gut-glow Chana Masala, served with brown rice
DINNER 'Feed-the-family' Rosemary Lentil Ragu	DINNER My Mama's Slow-cooked Jerk Chicken, Rice and Beans	DINNER Honey-mustard Chicken Thighs with Rosemary New Potatoes

NATURE'S PHARMACY: YOUR HERBAL-REMEDY GUIDE TO HORMONE BALANCE

Nature's pharmacy is a treasure trove of naturally healing and health-enhancing compounds to nourish your hormones to better health.

Please use herbal remedies with caution and consult your health practitioner if you are pregnant, breastfeeding, using hormone therapy or have an endocrine or mental-health disorder.

Traditionally, herbs have been used for generations for their medicinal powers to naturally balance levels of hormones in the body. While some may not have the scientific claims to back their hormone-balancing powers, I am going to introduce you to some evidence-based ones to give your hormones some extra support in combination with your hormone-optimised meals.

Vitex agnus-castus (chaste berry)

This is one of my hero herbal remedies when it comes to balancing hormones and regulating cycles. It encourages ovulation and increases progesterone during the luteal phase. It's an effective natural treatment used to support fertility, promote healthy, regular cycles and to treat symptoms of PCOS, PMDD, perimenopause and PMS.[130]

It stimulates ovulation by boosting dopamine and lowering prolactin levels, which helps to rebalance other hormones, including oestrogen and progesterone, and it has a calming effect on the nervous system.[131]

In one study, women with PMS took *vitex agnus-castus* during three consecutive menstrual cycles and 93 per cent of those reported a decrease in PMS symptoms (including depression and anxiety).

Another trial reported it to be superior to a placebo in reducing prolactin secretion, normalising a shortened luteal phase and increasing mid-luteal progesterone, all of which are vital for fertility.[132]

Ashwagandha

An apoptogenic herb – meaning it helps your body to deal with stress by moderating the brain's HPA axis – this is used to regulate cortisol, relax the mind and improve sleep. Studies suggest that ashwagandha supports adrenal function and may promote healthy DHEA levels (DHEA is an adrenal hormone necessary for energy, focus and vitality), so this herb may also be used to address symptoms of adrenal fatigue.

Red clover

This contains isoflavones, which are plant-based chemicals that produce oestrogen-like effects in the body. Isoflavones have shown potential in the treatment of conditions associated with menopause, such as hot flushes, as well as being supportive in cases of osteoporosis.[133] It also increases cervical mucus and lubricates the vagina to help tackle dryness, making it a great one in the lead-up to and during your ovulatory phase.

Dong quai

Sometimes called 'female ginseng', dong quai has been used in Chinese medicine for over 2,000 years to help balance hormones, reduce PMS symptoms and increase libido. It's also a circulatory stimulant, making it a helpful remedy for endometriosis, absent periods, dysmenorrhoea, scanty or unproductive periods.

These days, this herb is a common natural remedy for everything from the relief of menopausal symptoms to increased energy. It can also reduce inflammation, which is a benefit for those who have PCOS. A test-tube study showed that dong quai extract effectively decreased the levels of several different inflammatory markers.[134,135,136]

> ***Note:* *please consult your GP before taking dong quai if you are taking any medication.***

Maca

One study of post-menopausal women found that 3.5g per day of maca powder lowered measures of sexual dysfunction and decreased anxiety and depression symptoms after six weeks.[137] Also known as nature's HRT, maca is useful for improving other classic symptoms of menopause, like disrupted sleep, hot flushes and heart palpitations.

Rhodiola

An Ayurvedic herb, known for its ability to combat low mood by positively influencing levels of serotonin and dopamine in the brain, rhodiola is also a great tonic for burnout. One study noted increased happiness and significant improvements in fatigue and tension when supplementing with 400mg daily over several weeks.[138]

Sage (salvia)

Part of the mint family, this has been used for generations to reduce the impact of menopausal symptoms, including night sweats, hot flushes and mood swings. One study found that a fresh sage preparation lowered the severity and the number of hot flushes in menopausal women.[139] While sage is often taken as a tea, it's also available in capsule form and as an essential oil.

Shatavari

Known as the 'queen of herbs' and a fertility tonic, shatavari is used most often to support reproductive health.[140] It promotes oestrogen production, helps to make luteinising hormone and regulate cycles. This herb is also used in Ayurveda as an immunity booster, thanks to its impressive antioxidant profile. If you're trying to conceive, take this from menstruation until ovulation begins, then stop and resume next cycle. A typical dose of shatavari extract is 30 drops in water, up to three times daily. Avoid taking with other diuretic herbs or medication.

White peony

This has been shown to increase low progesterone, reduce high testosterone and regulate other hormones, including oestrogen and prolactin. But as well as balancing hormones, white peony also contributes to improved circulation in the pelvis area and better tone in the uterus, which help to build healthy blood flow to regulate menstruation. White peony is also particularly helpful for those with PCOS, PMS, bleeding in between periods and painful menstrual cramps, due to its analgesic and anti-inflammatory actions.

HORMONAL HARMONY HERBAL TEAS

Drinking herbal tea is one of the easiest ways to support all your body systems. It increases hydration levels, boosts your antioxidant intake, reduces inflammation, supports your gut microbiome and aids detoxification, all of which are essential for promoting hormonal harmony.

To make the teas below, use 1 teaspoon of dried herb leaves for every cup of water. Place the herbs in a cup, pour over boiling water and steep for 10 minutes. Strain before drinking.

Nettle tea

Nettle in the form of tea is a good way to support your cycle, especially a few days before, during and just after your period. This tea is a brilliant source of iron (to replenish lost stores in your menstrual blood) and magnesium and calcium (to reduce cramping and promote muscle relaxation).

Nettle is also a galactagogue – a herb that supports breast-milk supply, so this is one to drink during the fourth trimester of pregnancy.

> **TIP:** Just like all iron sources, adding a vitamin C source helps to aid absorption, so you could add a squeeze and a slice of lemon to your nettle tea.

Green tea

Green tea (and matcha) is high in protective plant compounds called polyphenols. The main bioactive compound, catechins and epigallocatechin gallate (EGCG), is known to support metabolism and improve insulin sensitivity in women with PCOS. It's also high in the amino acid L-theanine, associated with several health benefits, including improvements in mood, cognition and a reduction of stress and anxiety-like symptoms.[141]

Raspberry-leaf tea

This one is most known for its pregnancy benefits, particularly during the third trimester to facilitate labour and birth (I couldn't get enough of it in the weeks leading up to Sebby being born), and it's nice as an iced tea, too.[142] It's also a useful herbal remedy for the relief of PMS symptoms, largely down to its antioxidant profile, as it's high in compounds such as anthocyanins, which help to reduce inflammation-related symptoms, such as cramping and abdominal pain.[143] Raspberry-leaf tea is often referred to as a uterine tonic, meaning it may help tone and strengthen the uterine muscles; this is believed to reduce the intensity of menstrual cramps, making the pain associated with periods more manageable.

Spearmint tea

Spearmint tea has powerful anti-androgen effects in PCOS and research shows it can significantly decrease testosterone levels (which are elevated in PCOS) and increase levels of LH and FSH. These hormones play a vital role in regulating the menstrual cycle and improving egg quality. It's particularly beneficial to drink it during the follicular phase of the cycle.[144, 145]

Chamomile tea

Chamomile contains compounds that promote relaxation and sleep, among them its key bioactive antioxidant compound, apigenin, which binds to GABA receptors in the brain, reducing anxiety and promoting relaxation.[146] But it's also one of my absolute go-tos for soothing relief from PMS-associated symptoms, such as cramping (it is an anti-spasmodic and can ease tension in the uterine muscles) and a sensitive gut (it is a digestive relaxant and nourishing for your gut), in addition to anxiety, low mood and depression. What's more, it is antioxidant-rich and helps to reduce inflammation. And less inflammation equals being less prone to PMS.[147]

Ginger tea

Studies show that due to its anti-inflammatory effects, ginger is an effective herbal remedy for reducing PMS cramping[148] as it reduces the production of pain-causing prostaglandins and enhances pelvic circulation. One study found it just as effective as ibuprofen when it came to relieving menstrual pain.[149] Grate it into hot water or make a batch of my Immunity Tea Cubes (see page 170) for extra support, especially if you want to give your immune system some love.

Dandelion

Dandelion (*Taraxacum officinale*) is a powerful detoxifying and cleansing herb that neutralises toxins and facilitates the removal of waste products from the body via the liver, kidneys and gallbladder. Dandelion is also good for stimulating the bowels due to its mild laxative effect which helps eliminate toxins via stools. For hormonal acne and PMS, dandelion works by breaking down excess oestrogen and eliminating it from the body. When oestrogen builds up and is not excreted, it can cause a number of hormonal issues. Dandelion also exerts an anti-inflammatory action which can help with hormonal acne.

Milk thistle

Great for supporting natural detox pathways in the liver, milk thistle is effective for supporting clear skin from within. It contains antioxidants and a special phytonutrient called silymarin, which is the compound responsible for its detox benefits. In the liver, this apoptogenic herb helps to process and eliminate toxins, as well as metabolise excess hormones before they have a chance to recirculate and cause hormone-balance issues, like oestrogen dominance.

MEDICINAL MUSHROOMS

Apoptogenic mushrooms have proven health benefits, including cancer-fighting properties and brain-boosting compounds.

Medicinal mushrooms are a type of fungi that have been used for centuries in Eastern medicine to restore vitality and bolster health. Each has its own superpower, and here are some of my favourites that I use myself and in clinical practice.

Reishi – for stress relief, calm and sleep
Chaga – for energy
Cordyceps – for performance and endurance
Lion's mane – for cognition and focus
Tremella – for hair, skin and nails
Turkey tail – for immunity
Maitake mushroom - for blood sugar balance

Add the mushroom powders to hot water to make a tea or add them to your morning smoothie to supercharge it.

SUGGESTED HOLISTIC TREATMENTS

Hormone Testing

When it comes to testing, I cannot recommend the DUTCH (dried urine test for comprehensive hormones) highly enough. The most comprehensive hormone test out there, it shows how you process and metabolise your hormones (including progesterone, oestrogens and androgens), adrenal hormone metabolites (cortisol, cortisone, creatinine and DHEA-S), organic acids and melatonin metabolites.

Acupuncture

If I could recommend one holistic complementary therapy to invest your time and money in, alongside your optimised diet and lifestyle, it would be acupuncture. I had weekly sessions when I was at my lowest, most unbalanced self (when I first got my PCOS diagnosis).

Acupuncture has been shown to reduce hyperandrogenism (where there are high levels of androgens) and improve menstrual frequency in PCOS, and research supports its ability to positively influence sex hormones in gynaecological conditions.[150]

Acupuncture helped me not only to balance my hormones, but also to regulate my nervous system at a time when I was suffering with panic attacks and social anxiety. It helped to increase blood flow to my reproductive area, and I know it assisted us in our TTC (trying-to-conceive) journey. My tip would be to listen to a guided meditation during your treatment. You will sleep like a baby afterwards!

Infrared Saunas

Infrared saunas induce deep sweating, aiding in the elimination of toxins and waste products. This detoxification process eases the burden on your liver and helps maintain hormone balance.

Lymphatic Drainage

This treatment supports the body's natural detoxification processes and promotes hormone balance by eliminating toxins/used hormones in the body that need to be excreted. It promotes blood flow, reduces inflammation, supports the immune system and as a bonus it de-bloats like magic!

Reiki

Harmonises the hormonal system and stimulates the body's natural healing process. It's a treatment that instantly rebalances me and makes me feel renewed. My auntie Heidi is a reiki master, so I've been lucky enough to be treated by her regularly and I find it truly grounding.

NOTES

Introduction

1 https://www.ncbi.nlm.nih.gov/pmc/articles/PMC4443295/

Chapter 1: Eating Intuitively for Hormone Balance

2 https://www.ncbi.nlm.nih.gov/books/NBK547692/

3 https://www.naturalmedicinejournal.com/journal/ultra-processed-food-consumption-appetite-and-weight-gain#:~:text=This%20study%20confirms%20that%20a,peptide%20tyrosine%20tyrosine%20(PYY)

4 https://www.sciencedirect.com/science/article/abs/pii/S0301211510005671

5 https://www.ncbi.nlm.nih.gov/pmc/articles/PMC6160589/#:~:text=Conclusion%3A%20In%20obesity%2C%20circulating%20ghrelin,with%20normal%%20blood%20glucose

6. https://www.bbc.co.uk/news/health-18779997

7 https://www.ncbi.nlm.nih.gov/pmc/articles/PMC9145134/

8 https://www.ncbi.nlm.nih.gov/pmc/articles/PMC6160589/#:~:text=Conclusion%3A%20In%20obesity%2C%20circulating%20ghrelin,with%20normal%%20blood%20glucose

9 https://www.ncbi.nlm.nih.gov/pmc/articles/PMC6160589/#:~:text=Conclusion%3A%20In%20obesity%2C%20circulating%20ghrelin,with%20normal%%20blood%20glucose

10 https://pubmed.ncbi.nlm.nih.gov/19875483/

11 https://www.sciencedirect.com/science/article/pii/S0031938415300317

12 https://www.ncbi.nlm.nih.gov/pmc/articles/PMC4798912/#:~:text=The%20gut%2Dbrain%20axis%20is,function%20in%20health%20and%20disease

13 https://journals.asm.org/doi/10.1128/msystems.00031-18

14 https://zoe.com/learn/30-plants-per-week

Chapter 2: Getting to Know Your Hormones and the Reproductive System

15 https://pubmed.ncbi.nlm.nih.gov/18592262/

16 https://www.ncbi.nlm.nih.gov/books/NBK441996/

17 https://www.sciencedirect.com/topics/agricultural-and-biological-sciences/hormone-imbalance

18 https://pubmed.ncbi.nlm.nih.gov/23852908/#:~:text=It%20has%20been%20shown%20that,to%20women%20who%20do%20not

19.https://www.ncbi.nlm.nih.gov/pmc/articles/PMC3048776/

20 https://www.sciencedirect.com/topics/medicine-and-dentistry/indole-3-carbinol#:~:text=Indole%2D3%2DCarbinol%2FDiindolylmethane,detoxifying%20enzymes%20in%20the%20body

Chapter 3: Your Cycle-syncing Guide

21 https://www.sciencedirect.com/topics/agricultural-and-biological-sciences/hormone-imbalance

22 *Ibid.*

23 https://www.ncbi.nlm.nih.gov/pmc/articles/PMC4375225/

24 https://pubmed.ncbi.nlm.nih.gov/10682869/

25 https://www.ncbi.nlm.nih.gov/pmc/articles/PMC8599883/

26 https://www.ncbi.nlm.nih.gov/pmc/articles/PMC4488002/

27 https://www.ncbi.nlm.nih.gov/pmc/articles/PMC10261760/

28 *Ibid.*

29 *Ibid.*

30 https://www.ncbi.nlm.nih.gov/pmc/articles/PMC6429205/

31 https://pubmed.ncbi.nlm.nih.gov/8077314/

32 https://www.ncbi.nlm.nih.gov/pmc/articles/PMC10261760/

33 https://www.ncbi.nlm.nih.gov/pmc/articles/PMC10261760/#:~:text=Khanage%20(2019)%20concluded%20that%20the,et%20al.%2C%202021

34 https://www.ncbi.nlm.nih.gov/pmc/articles/PMC6079277/

35 https://www.ncbi.nlm.nih.gov/pmc/articles/PMC8634384/

36 https://ods.od.nih.gov/factsheets/VitaminB6-HealthProfessional/

37 *Ibid.*

38 https://www.ncbi.nlm.nih.gov/pmc/articles/PMC7414074/

39 https://pubmed.ncbi.nlm.nih.gov/26428278/

40 *Ibid.*

41 https://academic.oup.com/toxsci/article/176/2/253/5835885

Chapter 4: The Gut and Digestion

42 https://www.ncbi.nlm.nih.gov/pmc/articles/PMC9862683/

43 https://www.ncbi.nlm.nih.gov/pmc/articles/PMC2875955/

44 *Ibid.*

45 *Ibid.*

46 *Ibid.*

47 *Ibid.*

48 https://www.ncbi.nlm.nih.gov/pmc/articles/PMC4728667/

49 https://www.ncbi.nlm.nih.gov/pmc/articles/PMC4701845/

50 *Ibid.*

51 https://onlinelibrary.wiley.com/doi/10.1002/ijc.25207

52 https://pubmed.ncbi.nlm.nih.gov/34024716/

53 https://www.ncbi.nlm.nih.gov/pmc/articles/PMC8506209/

54 https://www.ncbi.nlm.nih.gov/pmc/articles/PMC6463098/

55 https://pubmed.ncbi.nlm.nih.gov/28778332/

56 https://pubmed.ncbi.nlm.nih.gov/7227774/

57 https://www.ncbi.nlm.nih.gov/pmc/articles/PMC3901893/

58 https://www.ncbi.nlm.nih.gov/pmc/articles/PMC5722595/

59 https://www.ncbi.nlm.nih.gov/pmc/articles/PMC4536296/

60 https://pubmed.ncbi.nlm.nih.gov/10470601/

61 https://www.ncbi.nlm.nih.gov/pmc/articles/PMC9572406

62 https://www.hopkinsmedicine.org/health/conditions-and-diseases/constipation

63 https://zoe.com/learn/prebiotic-foods

64 https://zoe.com/learn/best-probiotic-foods

65 https://www.healthline.com/health/phytonutrients#outlook

66 https://pubmed.ncbi.nlm.nih.gov/29568082/

67 https://www.ncbi.nlm.nih.gov/pmc/articles/PMC2875955/

68 *Ibid.*

69 https://www.hopkinsmedicine.org/health/conditionsand-diseases/constipation

70 https://www.ncbi.nlm.nih.gov/pmc/articles/PMC3093095/

71 https://www.ncbi.nlm.nih.gov/pmc/articles/PMC9592814/

72 https://www.ncbi.nlm.nih.gov/pmc/articles/PMC3850026/

Chapter 5: The Liver

73 https://www.ncbi.nlm.nih.gov/pmc/articles/PMC10200485/

74 https://www.ncbi.nlm.nih.gov/pmc/articles/PMC3399949/

76 https://pubmed.ncbi.nlm.nih.gov/32857150/

77 https://www.ncbi.nlm.nih.gov/pmc/articles/PMC8637678/#:~:text=levels%5B20%5D.-,Hormone%20metabolism,and%20inactivation%20of%20the%20hormones.

77 *Ibid.*

78 https://www.ncbi.nlm.nih.gov/pmc/articles/PMC7455852/#:~:text=Endocrine%2Ddisrupting%20chemicals%20are%20used,with%20behaviors%20characteristic%20of%20ADHD.

79 https://www.ncbi.nlm.nih.gov/pmc/articles/PMC5313342/

80 https://www.ncbi.nlm.nih.gov/pmc/articles/PMC10296738/

81 https://www.ncbi.nlm.nih.gov/pmc/articles/PMC2726844/

82 https://www.hormones-australia.org.au/the-endocrine-system/

83 https://www.ncbi.nlm.nih.gov/pmc/articles/PMC5657429/

84 https://www.niehs.nih.gov/health/topics/agents/endocrine#:~:text=Endocrine%2Ddisrupting%20chemicals%20(EDCs),part%20of%20the%20endocrine%20system.

85 https://www.ncbi.nlm.nih.gov/pmc/articles/PMC5657429/

86 https://www.healthline.com/nutrition/herbs-for-liver#1.-Milk-thistle-(silymarin)

87 https://pubmed.ncbi.nlm.nih.gov/2507689/

88 https://www.ncbi.nlm.nih.gov/pmc/articles/PMC3257687/

89 https://pubmed.ncbi.nlm.nih.gov/29389585/

Chapter 6: The Skin

90 https://pubmed.ncbi.nlm.nih.gov/23871029/

91 https://www.ncbi.nlm.nih.gov/pmc/articles/PMC10343488/#:~:text=By%20targeting%20estrogen%20receptors%2C%20quercetin,reproductive%20disorders%20associated%20with%20inflammation.

92 https://www.ncbi.nlm.nih.gov/pmc/articles/PMC5788264/

93 https://www.nccih.nih.gov/health/probiotics-what-you-need-to-know

94 https://www.ncbi.nlm.nih.gov/pmc/articles/PMC4488002/

95 https://pubmed.ncbi.nlm.nih.gov/30680163/

96 https://www.ncbi.nlm.nih.gov/pmc/articles/PMC7020168/

Chapter 7: The Neuroendocrine System

97 https://www.ncbi.nlm.nih.gov/pmc/articles/PMC3016669/

98 https://www.frontiersin.org/articles/10.3389/neuro.09.031.2009/full

99 https://www.ncbi.nlm.nih.gov/pmc/articles/PMC4316409/

100 https://www.sciencedirect.com/topics/psychology/neuroendocrine-system#:~:text=The%20neuroendocrine%20system%20is%20complex,affecting%20the%20hypothalamus%20%5B13%5D.

101 https://www.ncbi.nlm.nih.gov/pmc/articles/PMC6469458/

102 https://www.ncbi.nlm.nih.gov/pmc/articles/PMC6469458/

103 https://www.ncbi.nlm.nih.gov/pmc/articles/PMC3402070/#:~:text=Synthesis%20of%20melatonin%20requires%20tryptophan,compose%20serotonin%20and%20subsequently%20melatonin.

104 https://www.ncbi.nlm.nih.gov/pmc/articles/PMC6751071/

105 https://www.niddk.nih.gov/health-information/endocrine-diseases/national-hormone-pituitary-program/health-alert-adrenal-crisis-causes-death-people-treated-hgh

106 https://www.sciencedirect.com/science/article/pii/S2667325822001273

107 https://pubmed.ncbi.nlm.nih.gov/27870427/

108 https://www.ncbi.nlm.nih.gov/pmc/articles/PMC8190071/

109 https://pubmed.ncbi.nlm.nih.gov/31790663/

110 https://www.ncbi.nlm.nih.gov/pmc/articles/PMC3753111/

111 https://pubmed.ncbi.nlm.nih.gov/19499625/

112 https://www.sciencedirect.com/science/article/abs/pii/S0261561421002351#:~:text=Intake%20of%20simple%20carbohydrates%20('sugars,short%2D%20and%20long%2Dterm.

113 https://zoe.com/learn/gut-brain-connection

114 https://link.springer.com/article/10.1007/s00213-014-3810-0

115 https://zoe.com/learn/prebiotic-foods

116 https://www.ncbi.nlm.nih.gov/pmc/articles/PMC6769512/

1117 https://www.healthline.com/nutrition/foods-high-in-antioxidants#kale

118 https://pubmed.ncbi.nlm.nih.gov/35807749/

119 https://www.ncbi.nlm.nih.gov/pmc/articles/PMC5816734/

120 https://pubmed.ncbi.nlm.nih.gov/23642943/

121 https://pubmed.ncbi.nlm.nih.gov/35889834/

122 https://www.ncbi.nlm.nih.gov/pmc/articles/PMC6627761/

123 https://www.ncbi.nlm.nih.gov/pmc/articles/PMC6893582/#:~:text=The%20stress%20and%20cortisol%20reducing,62%2C74%2C75%5D.

Chapter 8: Putting it into Practice

124 https://www.medicalnewstoday.com/articles/320630

125 https://www.ncbi.nlm.nih.gov/pmc/articles/PMC6213178/

126 https://www.ncbi.nlm.nih.gov/pmc/articles/PMC7186155/#:~:text=Vitamins%20B2%2C%20B6%2C%20and%20B12%20are%20key%20players%20in%20one,possibly%20through%20affecting%20reproductive%20hormones.

127 https://www.ncbi.nlm.nih.gov/pmc/articles/PMC6996468/

128 https://www.healthline.com/nutrition/10-foods-high-in-magnesium#fa-q

129 https://www.ncbi.nlm.nih.gov/pmc/articles/PMC8850771/

130 https://pubmed.ncbi.nlm.nih.gov/23136064/

131 https://www.thieme-connect.de/products/ejournals/abstract/10.1055/s-0032-1327831

132 *Ibid.*

133 https://www.ncbi.nlm.nih.gov/pmc/articles/PMC8498057/

134 https://www.sciencedirect.com/topics/medicine-and-dentistry/angelica-sinensis

135 https://pubmed.ncbi.nlm.nih.gov/16691630/

136 https://pubmed.ncbi.nlm.nih.gov/21976127/

137 https://pubmed.ncbi.nlm.nih.gov/18784609/

138 https://www.ncbi.nlm.nih.gov/pmc/articles/PMC5370380/

139 https://link.springer.com/article/10.1007/s12325-011-0027-z

140 https://www.ncbi.nlm.nih.gov/pmc/articles/PMC4027291/

141 https://www.ncbi.nlm.nih.gov/pmc/articles/PMC5441188/

142 https://pubmed.ncbi.nlm.nih.gov/35188267/

143 https://www.ncbi.nlm.nih.gov/pmc/articles/PMC4931538/

144 https://pub med.ncbi.nlm.nih.gov/19585478/

145 https://pubmed.ncbi.nlm.nih.gov/17310494/

146 https://www.ncbi.nlm.nih.gov/pmc/articles/PMC6970572/#:~:text=The%20most%20common%20form%20is,of%20depressive%20symptoms%20%5B11%5D.

147 https://www.ncbi.nlm.nih.gov/pmc/articles/PMC6970572/#:~:text=The%20most%20common%20form%20is,of%20depressive%20symptoms%20%5B11%5D.

148 https://www.ncbi.nlm.nih.gov/pmc/articles/PMC4040198/

149 https://www.ncbi.nlm.nih.gov/pmc/articles/PMC4040198/

150 https://www.ncbi.nlm.nih.gov/pmc/articles/PMC3962314/

ACKNOWLEDGEMENTS

To my husband James, thank you for not just telling me every day, but showing me what true, unconditional love really is. Even in the hardest of times you have not only put up with me but supported every single one of my dreams (even the crazy ones!) and have been my rock, my home and my happiness. You have always made me feel anything is possible and I really love you for that.

Thank you for being the best father and role model to our little boy and enabling me to be the mother I've always dreamed of being, while building a career I am so proud of. I will cherish this privilege for ever. I could never have done any of this without you. I love you and I love us.

To my darling little boy Sebby, the light of my life. Wanting you more than anything I've ever wanted is what sparked my hormone-balance journey. None of this would be my reality without wanting you in the first place. Only good things came from wanting you. It was all for you. I am so proud of the loving and kind little boy you are, and I am so lucky that I get to be your mummy. I love you endlessly.

To my hero of a Dad, thank you for telling me since I was a little girl, 'If you think you can, you can'. Turns out you were very right. You have no idea how much your support and positivity rubbed-off on me and still does. And thank you for my Maltese heritage; good, home-cooked food runs in our blood and has made me the foodie I am today! I wish Nanny could see this book – I know it would make her smile, and probably cry.

To my beautiful Mum, thank you for showing me how to be a strong resilient woman with my own mind, I learnt from the best and I'm eternally grateful to you and Dad for instilling the core values in me that I am most proud of, and for showing me the definition of what hard graft and determination really is.

Thank you both for loving me and believing in me before I had the confidence to love and believe in myself.

To my baby brother George, thank you for calling me everyday just to see how I am. After 11 years as an only child you were the greatest gift that mum and dad could've given me. Siblings by blood, best friends by choice.

To Karen and Paul for loving me like your own daughter and for letting me use your kitchen to test my recipes. Mummy Shand – we finally perfected the granola and seed cycling crackers!! thank you for always being so patient with me.

To Amy (Bister), no one has the ability to make me laugh like you do, even at the most hormonal of times you've always known what to say to turn things around and make me smile. Thank you for being by my side every step of the way in writing this book and in every area of my life. I love you like family and I am so grateful I get to do life (and work) with you.

To Natalie (Twinny), thank you for being my cheerleader from the beginning of this whole process, I love how much you wanted this for me. You were the friend I didn't know I needed at a time in my life I needed you the most. Life wouldn't be life without you, Beebs, Mark and Reggie in it.

To Hannah, no matter how far away we are or how long we've been apart, being with you always feels like home. I will forever cherish you and our special friendship. Thank you for always encouraging me in all that I do. I hope you know how lucky I feel to have you in my life.

To Sarita (Sis), thank you for always being there to laugh, dance and cry with. I have always felt so supported by you. I love the core memories we've made together and the ones I know we will make. Adore you so much, you special soul.

To Beth, thank you for all of the dog walks and life chats, we've walked, chatted (and cried) through every season of our adult lives. Thank you for always giving me such grounding advice on all the things.

To Hayley, I'm so grateful I met you that day on Epping High Street, you've been in mine and Sebby's life every day since and you are stuck with us forever now! I love that we leave batches of home-cooked food on each other's doorsteps for our little P's. I hope we always do this.

To Melissa (Melly), thank you for everything you have helped me with in building Eat, Nourish and Glow. You really are one of a kind and I love the friendship that has come from working together.

To my incredible editor Lydia for making my ultimate vision board goal come to life and giving me the opportunity to write the book I manifested writing, thank you so much I'll be forever grateful that the universe bought us together. Thank you to Sarah, Catherine, Rosie, Lizzie and the rest of the talented HarperCollins team for all of your hard work on my book. I am completely and utterly blown away by it, I pinch myself every time I pick it up.

To each and every client I've had the privilege of treating, I have and will continue to learn so much from you. You are what makes me want to get up in the morning and do a job I want to spend the rest of my days doing. Thank you for trusting me with your health, it's truly the biggest honour.

To my loyal and lovely online community, I can't tell you just how much I appreciate your continuous support. Thank you from the bottom of my heart for making my recipes and showing your kindness in every share, comment, like and DM, nothing has gone unnoticed. You have enabled me to share my hormone balance mission and this book wouldn't be here without you.

INDEX

This book contains advice and information relating to health care. It should be used to supplement rather than replace the advice of your doctor or another trained health professional. If you know or suspect you have a health problem, it is recommended that you seek your doctor's advice before embarking on any medical programme or treatment. All efforts have been made to assure the accuracy of the information contained in this book as of the date of publication. This publisher and the author disclaim liability for any medical outcomes that may occur as a result of applying the methods suggested in this book. When using kitchen appliances please always follow the manufacturer's instructions.

Thorsons
An imprint of HarperCollins*Publishers*
1 London Bridge Street
London SE1 9GF

www.harpercollins.co.uk

HarperCollins*Publishers*
Macken House, 39/40 Mayor Street Upper
Dublin 1, D01 C9W8, Ireland

First published by Thorsons 2025

10 9 8 7 6 5 4 3 2 1

A catalogue record of this book is available from the British Library

ISBN 978-0-00-869824-9

Food Stylist: Rosie Reynolds
Prop Stylist: Louie Waller

Printed and bound by GPS Group, Bosnia-Herzegovinia

This book contains FSC™ certified paper and other controlled sources to ensure responsible forest management.

For more information visit: www.harpercollins.co.uk/green